MEASURING PEACE

Praise for *Measuring Peace*

'Caplan is a leading scholar in the effort to encourage the international community to take the measurement of peace more seriously.'

Andrew Rathmell, *The RUSI Journal*

'This book...provides a much needed framework for policy deliberations that might eventually contribute toward preventing the recurrence of conflicts.'

Abhishek Choudhary, *Peace & Change*

'*Measuring Peace* is a spectacular scholarly achievement and clearly shows where academia can have a policy impact. It can serve as a useful tool in the hands of peacebuilders facing the daunting task of measuring the quality of peace.'

Jessie Barton Hronešová, *OXPOL: The Oxford University Politics Blog*

'*Measuring Peace* adds significantly and provocatively to a fractious and controversial but nevertheless enduring conversation about what peace is—conceptually, operationally and practically—as well as considering how we should study this crucial phenomenon.'

Christian Davenport, *Perspectives on Politics*

'Caplan's core argument is that more rigorous assessments of peace are needed. In his findings he emphasizes a number of principles of good practice that can contribute more rigour to the assessment of peacebuilding initiatives.'

Cedric de Coning, *Ethnopolitics*

Measuring Peace

Principles, Practices, and Politics

RICHARD CAPLAN

OXFORD
UNIVERSITY PRESS

OXFORD
UNIVERSITY PRESS

Great Clarendon Street, Oxford, OX2 6DP,
United Kingdom

Oxford University Press is a department of the University of Oxford.
It furthers the University's objective of excellence in research, scholarship,
and education by publishing worldwide. Oxford is a registered trade mark of
Oxford University Press in the UK and in certain other countries

© Richard Caplan 2019

The moral rights of the author have been asserted

First published 2019
First published in paperback 2021

All rights reserved. No part of this publication may be reproduced, stored in
a retrieval system, or transmitted, in any form or by any means, without the
prior permission in writing of Oxford University Press, or as expressly permitted
by law, by licence or under terms agreed with the appropriate reprographics
rights organization. Enquiries concerning reproduction outside the scope of the
above should be sent to the Rights Department, Oxford University Press, at the
address above

You must not circulate this work in any other form
and you must impose this same condition on any acquirer

Published in the United States of America by Oxford University Press
198 Madison Avenue, New York, NY 10016, United States of America

British Library Cataloguing in Publication Data
Data available

Library of Congress Cataloging in Publication Data
Data available

ISBN 978–0–19–881036–0 (Hbk.)
ISBN 978–0–19–886770–8 (Pbk.)

Links to third party websites are provided by Oxford in good faith and
for information only. Oxford disclaims any responsibility for the materials
contained in any third party website referenced in this work.

For Daniel

Acknowledgements

This book has been (too) long in the making, and I am very grateful to many individuals and institutions for their valuable input and support along the way.

I owe a very large debt of gratitude to the Folke Bernadotte Academy, which took a keen interest in this project from its inception and made significant contributions to it, intellectual as well as financial, from beginning to end. The Folke Bernadotte Academy's commitment to scholarly research in the fields of peace- and state-building is truly impressive.

Without the benefit of a British Academy fellowship, I would never have found the time to undertake the initial research that started the ball rolling. I am grateful to the Academy for taking a gamble on me.

The UK Department for International Development was very generous in supporting the research that underpins Chapter 4 of this book. They also took an active interest in the research itself and made very valuable contributions to it.

Many of the ideas for this project originated with a consultancy I did for the United Nations (UN) Peacebuilding Support Office (PBSO) a number of years ago on 'measuring peace consolidation and managing transitions', which led to the production of an internal briefing paper that I had the honour to present to the UN Peacebuilding Commission. I was pleased that one of my recommendations—for the production of a benchmarking handbook—was taken up by the PBSO, resulting in the publication of *Monitoring Peace Consolidation: United Nations Practitioners' Guide to Benchmarking*, expertly written by Sven Erik Stave of Fafo. It has been exciting to see greater and more effective use of benchmarking as an instrument of assessment throughout the UN system, as I discuss in Chapter 3.

The ideas that emerged from my work with PBSO were explored further in the context of two conferences I organized that brought together scholars and practitioners at Wilton Park, the global forum based at Wiston House in the bucolic British countryside of West Sussex. The first conference, on exit strategies and peace consolidation in state-building operations, was held on 13–15 March 2009. The second conference, on measuring peace consolidation, was held on

viii   *Acknowledgements*

15–17 October 2014. Both conferences allowed for an unfettered
exchange of ideas and experiences. I am grateful to Wilton Park
and its programme director Isobelle Jaques, in particular, for their
assistance in organizing the conferences. Financial support was provided by the Centre for International Studies, University of Oxford;
the Fafo Institute for Applied International Studies; the Norwegian
Peacebuilding Resource Centre; the Folke Bernadotte Academy; the
Swiss Federal Department of Foreign Affairs; the British Academy;
the Organization for Security and Co-operation in Europe, Vienna;
and the Public Diplomacy Division, North Atlantic Treaty Organization, Brussels. I am grateful to all of them for their generous
assistance.

I am also grateful to Linacre College Oxford for the grant I received
from the Lucy Halsall Fund in support of my overseas research
expenses.

For their very valuable research assistance, I am grateful to Nicholas
Barker, Kate Brooks, and Allard Duursma.

A number of individuals read all or parts of this manuscript and
I am extremely grateful to them for their input: Nicholas Barker, Alex
Bellamy, Jane Boulden, Frances Brown, John Gledhill, Anke Hoeffler,
Lucas Kello, Lara Olson, Michael von der Schulenburg, and Remco
Zwetsloot. I have also benefitted from the feedback I received from
presentations to the Oxford International Relations Colloquium and
the Oxford University Strategic Studies Group, and from conversa-
tions with many of my Oxford colleagues.

A version of the Introduction to this book appeared as 'Measuring
Peace Consolidation' in the British Academy's *Rethinking State Fra-
gility* (London: British Academy, 2015). I am grateful to the British
Academy for granting me permission to use this material.

Chapter 4, co-written with Anke Hoeffler, appeared in modified
form in the *European Journal of International Security* (Vol. 2, No. 2,
July 2017). We are grateful to Cambridge University Press for granting us permission to adapt this article for inclusion in this volume.
We are also grateful to Lise Howard for the use of her UN peacekeeping operations data and to Kate Roll for updating it. Chris Perry
gave very helpful advice on the use of the International Peace Institute
data on UN peacekeeping. Joakim Kreutz clarified the use of the
UCDP conflict termination data. The FHI 360 Education Policy
and Data Center provided data on horizontal inequality. Daniel
Gutknecht, Ron Smith, Måns Söderbom, the six case study authors,

Acknowledgements ix

and the participants in the project meeting in Oxford on 6 February 2015 all provided useful comments and suggestions.

Dominic Byatt at Oxford University Press has been as patient and encouraging an editor as one could ever hope to have. I am grateful for his support of this and earlier projects of mine. Olivia Wells provided superb editorial assistance.

Finally, I thank my wife Luisa, for her forbearance especially, and my son Daniel, to whom this book is dedicated. May he and his generation know more peaceful times.

Contents

List of Figures	xiii
List of Tables	xv
List of Abbreviations	xvii
Introduction	1
1. Conceptualizing Peace	13
2. From Conception to Practice	30
3. Assessing Progress	51
4. Factors of Post-Conflict Peace Stabilization *With Anke Hoeffler*	77
5. Measuring Peace Consolidation	104
Conclusion	123
Select Bibliography	127
Index	145

List of Figures

1.1. Total armed conflict by type, 1946–2014	16
3.1. ISAF notional assessment summary slide for one campaign task	65
4.1. Kaplan-Meier survival estimate	85
4.2. Kaplan-Meier survival estimates	86
4.3. Kaplan-Meier survival estimates	87

List of Tables

0.1.	Civil war onset and recurrence	3
1.1.	Core features of assessing the state of peace or rivalry between countries	23
4.1.	Armed conflict outcomes, 1990–2013	84
4.2.	Number of peace spells surviving	85
4.3.	Duration of peace and past conflict characteristics	90
4.4.	Duration of peace: territorial and ethnic conflicts and income	92
4.5.	Duration of peace and UNPKOs	95
4.6.	UN peacekeeping operations	96
4.7.	UNPKOs and peace settlements (case studies)	102

List of Abbreviations

ACCORD	African Centre for the Constructive Resolution of Disputes
ACD	Armed Conflict Dataset
AU	African Union
BNUB	United Nations Office in Burundi
CAR	Central African Republic
COW	Correlates of War
DDR	disarmament, demobilization, and reintegration
DPA	Department of Political Affairs
DPKO	Department of Peacekeeping Operations
DRC	Democratic Republic of Congo
ECOWAS	Economic Community of West African States
FSI	Fragile/Failed States Index
IEP	Institute for Economics and Peace
IMF	International Monetary Fund
ISAC	ISAF Strategic Assessment Capability
ISAF	International Security Assistance Force
NATO	North Atlantic Treaty Organization
NGO	non-governmental organization
OECD	Organisation for Economic Co-operation and Development
OSCE	Organization for Security and Co-operation in Europe
PBC	Peacebuilding Commission
PBSO	Peacebuilding Support Office
PCPI	Post-Conflict Performance Indicator
PCRD	Post-Conflict Reconstruction and Development
PPI	Positive Peace Index
PRIO	Peace Research Institute, Oslo
PRSP	Poverty Reduction Strategy Paper
PSC	Peace and Security Council
UCDP	Uppsala Conflict Data Program
UN	United Nations
UNAMSIL	United Nations Mission in Sierra Leone
UNMIL	United Nations Mission in Liberia
UNPKO	United Nations Peacekeeping Operation
USAID	United States Agency for International Development

Introduction

How can we know if the peace that has been established following a civil war is a stable peace?

Much hinges on the answer to this question. Each year intergovernmental organizations, donor governments, and non-governmental organizations expend billions of dollars and deploy tens of thousands of personnel in support of efforts to build peace in countries emerging from violent conflict. The United Nations (UN) alone at the end of 2017 had nearly 93,000 uniformed personnel in the field and was slated to spend some $6.8 billion on peacekeeping operations in that financial year.[1] Yet despite this considerable commitment of resources, as well as the accumulation of extensive knowledge and experience relevant to peacebuilding in the course of the past two decades, external efforts to consolidate peace in conflict-affected countries have met with mixed results.

The recurrence of violence in the Central African Republic (CAR) in late 2012 is a case in point. CAR is one of six countries on the agenda of the UN's Peacebuilding Commission, the UN body established in 2005 with a mandate to support recovery efforts in countries emerging from violent conflict.[2] Civil war raged in CAR from 2004 to 2007 until a peace agreement, an amnesty, and the formation of a national unity government laid the foundations for a durable peace,

---

[1] United Nations, 'Monthly Summary of Military and Police Contribution to United Nations Operations', 31 December 2017, https://peacekeeping.un.org/sites/default/files/msr_31_dec_2017_0.pdf. For my use of the terms 'peacebuilding', 'peacekeeping', and related terms, see the terminology section at the end of this chapter.

[2] UN Security Council Resolution 1645 (2005) and UN General Assembly Resolution 60/180 (2005), adopted concurrently on 20 December 2005, authorized the establishment of the Peacebuilding Commission. CAR was put on the agenda of the Peacebuilding Commission in 2008 at the request of the Bozizé government.

Measuring Peace

which the UN took measures to reinforce. Violent conflict re-erupted after rebel forces, accusing the government of François Bozizé of failing to abide by its commitments, staged a coup in December 2012. The fact that CAR suffered renewed armed hostilities on the UN's watch underscores the volatility of so-called post-conflict countries and the need to understand why peace may fail to consolidate despite substantial international engagement.[3]

CAR is not an isolated case. Between 1946 and 2013, 105 countries suffered civil wars of various magnitude. Of these, more than half (fifty-nine countries) experienced a relapse into violent conflict—in some cases more than once—after peace had been established.[4] By one estimate, on average 40 per cent of countries emerging from civil war are likely to revert to violent conflict within a decade of the cessation of hostilities.[5] According to the World Bank, 90 per cent of all civil wars that erupted in the first decade of the twenty-first century were in countries that had previously experienced a civil war since 1945 (see Table 0.1).[6] Many of these countries have been recipients of extensive post-conflict recovery assistance on the part of the international community.

Peace may fail for a variety of reasons, as we discuss below, but many efforts to build peace have been hampered in one important respect: by the lack of effective means of assessing progress towards the achievement of a consolidated peace. As a consequence, peacebuilders are often navigating without a compass. International organizations and donor governments routinely undertake monitoring and evaluation of the specific programmes that they support in countries recovering from violent conflict, often to determine if funds are being

---

[3] For an assessment of the situation on the eve of renewed hostilities, see 'Report of the UN Secretary-General on the Situation in the Central African Republic and on the Activities of the United Nations Integrated Peacebuilding Office in That Country', UN Doc. S/2012/956, 21 December 2012.

[4] Uppsala Conflict Data Program and Peace Research Institute, Oslo, 'UCDP/ PRIO Armed Conflict Dataset v.4-2014a, 1946–2013'.

[5] Paul Collier, Anke Hoeffler, and Måns Söderbom, 'Post-Conflict Risks', *Journal of Peace Research* 45:4 (2008), 465. Different studies yield different estimates of conflict relapse depending on the data, criteria, and methodology employed. These differences are not significant for the purposes of the analysis presented in this book, however. For a critical discussion of the varying estimates, see Astri Suhrke and Ingrid Samset, 'What's in a Figure? Estimating Recurrence of Civil War', *International Peacekeeping* 14:2 (2007), 195–203.

[6] World Bank, *World Development Report 2011: Conflict, Security and Development* (Washington, DC: World Bank, 2011), 3.

Introduction

Table 0.1. Civil war onset and recurrence

Decade	Onset in countries with no previous conflicts (%)	Onset in countries with a previous conflict (%)	Number of onsets
1960s	57	43	35
1970s	43	57	44
1980s	38	62	39
1990s	33	67	81
2000s	10	90	39

Source: World Bank 2011

used as intended or if programme activities have been implemented as planned. Rarely, if ever, however, do these organizations and governments conduct broader, strategic assessments to ascertain the quality of the peace that they are helping to build and the contribution that their engagement is making (or not) to the consolidation of peace.

This is not to suggest that peacebuilding actors make no effort to take stock of progress overall. To the contrary, there are periodic reports from the field by high representatives and their equivalents, briefings to organizations' member states and government ministers, and expert independent analysis by research institutes, among other barometers of change. While these assessments can be very insightful, they are often ad hoc, impressionistic, or devised on the basis of either inexplicit criteria or stated criteria—such as the fulfilment of mandates—that are not necessarily suitable for determining how well a peacebuilding operation may be helping to meet the requirements for a stable peace.

The key issue to consider, then, which this book will address, is can we know—and if so, how can we know—if the foundations for sustainable peace and development have been established so that the UN and other multilateral organizations, donor governments, and non-governmental organizations engaged in peacebuilding can decide whether, when, and in what ways they can recalibrate their engagement in these countries. While decisions of this kind will always be political ones ultimately,[7] a greater appreciation of the quality of the peace that has been established would arguably enable

---

[7] For examples, see Richard Caplan, 'Policy Implications', in Richard Caplan (ed.), *Exit Strategies and State Building* (New York: Oxford University Press, 2012), 315–16.

4   *Measuring Peace*

international actors engaged in post-conflict recovery and development
to make better informed judgements about appropriate courses of
action. To build a secure peace, it will be argued here, it is important
to take the measure of peace.

MEASURING PEACE CONSOLIDATION

How can one assess the durability of a peace? The principal difficulty
in attempting to answer this question is that there are no hard
measures or indicators of a consolidated peace—in contrast, say, to
the indicators of a prosperous economy (e.g., growth in gross domestic product) or a healthy population (e.g., declining infant mortality
rates), contentious though some of these indicators may be.[8] The
ultimate test of a sustainable peace, in cases where third parties have
intervened, necessarily comes after the fact—that is, only when the
international community has drawn down significantly or exited.
This difficulty is compounded by the fact that the continued presence
of international personnel, even just a token military presence, can
buoy a peace artificially. The presence of UN peacekeeping forces in
Liberia in 1997, for example, helped to keep the peace but it also gave
rise to mistaken impressions of the rootedness of that peace, which
the resumption of civil war less than two years later would dispel.[9]
One measure of sustainability, therefore, it has been suggested, is the
survival of a peace following the first election *after* peacekeeping
forces have departed.[10] Yet while this is conceivably a reasonable
measure, it is obviously not a practical one for transitional planning
purposes. Third parties want to know that a peace is stable *before*
they exit.

---

[8] See, for instance, the trenchant critique of gross domestic product as a measure
of economic well-being in Joseph E. Stiglitz, Amartya Sen, and Jean-Paul Fitoussi,
*Report by the Commission on the Measurement of Economic Performance and Social
Progress* (2009), http://ec.europa.eu/eurostat/documents/118025/118123/Fitoussi+
Commission+report.

[9] See 'Final Report of the Secretary-General on the United Nations Observer
Mission in Liberia', UN Doc. S/1997/712, 12 September 1997.

[10] Barry Blechman, William J. Durch, Wendy Eaton, and Julie Werbel, *Effective
Transitions from Peace Operations to Sustainable Peace: Final Report* (Washington,
DC: DFI International, September 1997), 8–9.

Introduction 5

There is a substantial body of scholarship concerned with civil wars and peace maintenance but that scholarship offers only limited insight into whether a post-conflict peace is durable. One area of scholarship with apparent relevance to this question is concerned with civil war onset. Scholars have identified a wide range of factors in their efforts to explain the incidence of violent internal conflict. Many of these factors can be grouped in terms of their primary emphasis: on the *motivation* of combatants and their supporters, on the *feasibility* of rebellion, and on the *resilience* of national institutions. Motivation encompasses a wide range of often grievance-based sub-factors, including 'relative deprivation' (Gurr 1970) and 'horizontal inequalities' (Stewart 2008); ethnic insecurity (Posen 1993; Walter and Snyder 1999); and political, social, and economic discrimination (Brown 1996). Feasibility stresses the importance of opportunity over motivation, suggesting that rebellion is more likely to occur where material conditions favour it, notably where the terrain is mountainous, allowing rebels to hide; where valuable natural resources are plentiful, allowing rebels to finance their activities from trade; and where external security commitments to governments are weak, allowing rebels to challenge governments more easily (Fearon and Laitin 2003; Collier et al. 2009). Resilience emphasizes the vulnerability of the state to various internal and external pressures (e.g., rising food prices, migration) and the capacity of states and their institutions to cope effectively with these pressures (Zartman 1995; Goldstone et al. 2010; World Bank 2011). These factors are not necessarily mutually exclusive: a number of explanations for the outbreak of civil war combine several of them.[11]

If one can identify the factors that underlie civil wars, it seems reasonable to assume, then the basis for an enduring peace will arguably reside in being able to address those factors satisfactorily—by eliminating discrimination, for instance, or by building more representative institutions—bearing in mind the difficulty of effecting some of these changes. There are two problems with this approach. The first problem is that there is no consensus among scholars as to which causal factors matter or matter most. Indeed, as Charles Call observes, there is 'tremendous disparity among scholars about

---

[11] For a review of the literature on civil war causation produced in the most recent period of scholarship, see Lars-Erik Cederman and Manuel Vogt, 'Dynamics and Logics of Civil War', *Journal of Conflict Resolution* 61:9 (2017), 1992–2016.

whether certain factors are important or not, and about the degree to which they are important'.[12] The identification of critical factors alone, moreover, is not sufficient to account for why conflict occurs; there needs also to be a credible and verifiable explanation of why they matter, and scholars disagree about that, too. For instance, scholars who agree that peacekeeping makes a positive contribution to peacebuilding maintain variously that it succeeds because it mitigates the security dilemma among warring parties (Fortna 2004); or because it reinforces negotiated settlements (Caplan and Hoeffler 2017); or because it constitutes a projection of soft power (Howard 2019). These differences matter for peacebuilding strategies.

The second problem with this approach to measuring peace consolidation is that it assumes that the causes of conflict onset and the causes of conflict recurrence are one and the same. Call's quantitative analysis has shown, however, that while onset and recurrence share a number of risk factors—including political instability, population size, and reliance on natural resource (notably oil) exports—there are also significant differences.[13] For one thing, 'wars are transformative', as Susan Woodward has observed; the root causes of a conflict may no longer pertain as a consequence of changes that the conflict may have generated—changes that include major population displacements and the emergence of new political or military elites.[14] It is important, therefore, to treat civil war recurrence—and the factors that give rise to it—as distinct phenomena.[15]

Another possible approach is to focus not on the causes of civil war but on the causes of peace in the aftermath of civil war. What measures have been most successful in maintaining the peace after violent conflict and have they been applied to the cases in question? Again, the range of possibilities—and the differences among scholars—is considerable. Scholars have stressed the importance of the nature of civil war terminations (Licklider 1993), third-party security guarantees (Fortna 2004), security-sector reform (Toft

---

[12] Charles T. Call, *Why Peace Fails: The Causes and Prevention of Civil War Recurrence* (Washington, DC: Georgetown University Press, 2012), 30.

[13] Ibid., ch. 2.

[14] Susan L. Woodward, 'Do the Root Causes of Civil War Matter? On Using Knowledge to Improve Peacebuilding Interventions', *Journal of Intervention and Statebuilding* 1:2 (2007), 155.

[15] Call, *Why Peace Fails*, 50–9, 65.

Introduction 7

2010), and inclusive political settlements (Call 2012), among other measures.[16] This approach would appear to be more promising in so far as it draws its analysis from experiences of success. But as with the previous approach, there is a lack of consensus among scholars and, most important for our purposes, this approach does not reveal enough about the quality of the peace that has been established in any given case.

This book proposes a different approach. The argument made here—a very simple argument—is that more rigorous assessments of the robustness of peace are needed. These assessments require clarity about the characteristics of, and the requirements for, a stable peace in a given conflict situation and correspondingly strong know-ledge about the conflict dynamics specific to that conflict situation. The objectives (intended outcomes) of a peacebuilding operation need to be re-evaluated continually. Do these objectives still support the broad strategic goals of the operation? Are the assumptions that underpin those objectives valid? Have new or unanticipated threats or impediments to a stable peace emerged (e.g., external security challenges, new political developments) that require the articulation of new or altered objectives? Has available implementing capacity—internationally and nationally—changed and what implications does this have for achieving a stable peace?[17] Such assessments are feasible; indeed, as we will see, they are being employed already by some peacebuilding bodies but only to a limited extent. More rigorous assessments of the robustness of peace, while by no means a panacea for conflict recurrence, have the potential to make substantial contributions to conflict prevention.

ORGANIZATION OF THE BOOK

This is a book about measuring peace consolidation. It is not a book about evaluating peacebuilding success, on which there is

[16] In fairness, not all of these scholars have been concerned with whether the measures in question have been the most effective but, rather, with how effective they have been.

[17] These same considerations, I argue elsewhere, ought to inform transitional planning for peace operations. See Caplan, *Exit Strategies and State Building*, ch. 17.

8   *Measuring Peace*

considerable scholarship.[18] The two topics are closely related but they are distinct. The first topic—the topic of this book—is concerned with assessing the quality of peace; the second topic is concerned with assessing peacebuilding performance. One measure of peacebuilding performance may be the quality of the peace that it produces, and in that sense the two topics are related, but it should be clear that they are distinct.

This book is organized around five chapters. Chapter 1 examines the concept at the heart of the book: peace. Every peacebuilding strategy is predicated on a conception of peace, whether implicit or explicit. It may be as basic as the absence of armed conflict—what is known as a 'negative peace'—or it may envision a more ambitious outcome such as the reconciliation of warring parties and the restoration of trust within war-torn societies—a 'positive peace'. This chapter will examine the range of conceptions of peace that have been proposed by scholars. It will argue that peace is more varied and heterogeneous a concept than either the scholarly literature or the policy literature often acknowledges. Conceptualizations that reflect the degrees of fragility/robustness of peace in post-conflict environments can provide the basis for sounder peacebuilding strategies.

Different conceptions of peace have different implications for devising strategies of peacebuilding and peace maintenance. What it takes to achieve a negative peace is very different from what is required to achieve a positive peace. Chapter 2 explores how the conceptual distinctions discussed in the previous chapter map onto actual practice, with reference to the principal multilateral actors engaged in peacebuilding: the UN, the Organization for Security and Co-operation in Europe, the North Atlantic Treaty Organization, the African Union, the World Bank, and leading non-governmental

---

[18] Relevant works include Duane Bratt, 'Assessing the Success of UN Peacekeeping Operations', *International Peacekeeping* 3:4 (1996), 64–81; Charles T. Call, 'Knowing Peace When You See It: Setting Standards for Peacebuilding Success', *Civil Wars* 10:2 (2008), 173–94; Paul F. Diehl and Daniel Druckman, 'Evaluating Peace Operations', in Joachim A. Koops, Noorie MacQueen, Thierry Tardy, and Paul D. Williams (eds), *The Oxford Handbook of United Nations Peacekeeping Operations* (Oxford: Oxford University Press, 2015), ch. 5; Michael W. Doyle and Nicholas Sambanis, *Making War and Building Peace: United Nations Peace Operations* (Princeton, NJ: Princeton University Press, 2006); Lise Morjé Howard, *UN Peacekeeping in Civil Wars* (New York: Cambridge University Press, 2008); Virginia Page Fortna, *Does Peacekeeping Work? Shaping Belligerents' Choices after Civil War* (Princeton, NJ: Princeton University Press, 2008).

Introduction 9

organizations. What are the primary features of these organizations' approaches to peacebuilding? How do they differ, if at all, in their understandings of the characteristics of, and requirements for, a stable peace?

Chapter 3 examines how international peacebuilding actors assess progress towards peace consolidation, to the extent that they do. Assessments are conducted both informally, for instance through the periodic reporting of heads of missions and briefings to organizations' member states and government ministers, and more formally, through benchmarking, conflict analysis, and early warning indicators, among other practices. The chapter highlights innovative approaches to strategic assessment that have yielded insights into the robustness of peace in specific cases. The chapter also examines some of the many indices and indicators of peace, stability, resilience, and the like that are produced periodically by think tanks and research institutes, including the Global Peace Index, the Peace and Conflict Instability Ledger, the Failed (now Fragile) States Index, and the Everyday Peace Indicators. For the most part, I argue, these indices conceal more than they reveal about the quality of post-conflict peace, but there are notable exceptions.

Chapter 4 draws on research that I have conducted jointly with my colleague Anke Hoeffler, which seeks to identify factors that contribute to post-conflict peace stabilization. The research has two main components: a quantitative analysis using duration (survival) analysis, and a qualitative analysis examining the peace consolidation process in six conflict-affected countries. Duration analysis, a statistical method, allows us to analyse the duration of peace. The hazard rate—the rate at which peace ends—can be modelled as a function of various co-variates, such as economic growth, aid, elections, military personnel and expenditure, regional autonomy, etc. The data for this purpose come from a wide range of sources, including data from UN sources that has not previously been available to researchers. The country case studies provide more detailed information on how some countries achieved lasting peace while others failed. The country cases that are included in this analysis are: Burundi, El Salvador, Liberia, Nepal, Sierra Leone, and Timor-Leste (East Timor).

Chapter 5 discusses the implications of the analysis in the foregoing chapters for international policy. What does this analysis tell us about whether it may be possible, and if so how, to assess the quality of peace? How can monitoring and assessment be improved? The

10 *Measuring Peace*

chapter argues for an 'ethnographic approach' to strategic assessment that favours increased reliance on knowledge of local culture, local history, and especially, the specific conflict dynamics at work in a given conflict situation, particularly at the micro level. From this approach it derives a number of recommended practices, including early and continuous conflict analysis, the adoption of more precise measures of post-conflict peace, and the incorporation of local perspectives into strategic assessment. The chapter closes with a discussion of the obstacles to good practice (e.g., politicization of reporting) and how these obstacles can be overcome.

A FEW WORDS ABOUT TERMINOLOGY

A few words about the terminology employed in this book are in order. The notion of a stable peace (also referred to in this volume as a secure, self-sustaining, robust, or consolidated peace) lies at the heart of this study. *Peace*, as both a policy goal and as an object of academic study, is an 'essentially contested concept' in W. B. Gallie's sense of the term: 'concepts, the proper use of which inevitably involves endless disputes about their proper uses on the part of their users'.[19] Is peace characterized by the absence (total? partial?) of armed conflict (defined how?) or does it exhibit (require?) additional features such as broad shifts (elite? popular?) in attitude? There is no consensus among scholars and practitioners as to the characteristics of peace. This is because ambiguity is inherent in the concept of peace.[20] Different appreciations of the concept, however—including from within conflict-affected communities—can have implications for both analysis and policy.

For the purposes of this volume, a stable peace is understood to mean a condition in which the recurrence of civil war is thought to be unlikely. Where it is necessary to define a stable peace more precisely (e.g., in Chapter 4), I do so; otherwise greater precision is

---

[19] W. B. Gallie, 'Essentially Contested Concepts', *Proceedings of the Aristotelian Society* 56 (1956), 169.

[20] Michael Lipson, 'Performance under Ambiguity: International Organization Performance in UN Peacekeeping', *Review of International Organizations* 5:3 (2010), 249.

Introduction 11

not necessary but an awareness of the different meanings of the term and their implications is. The notion of a stable peace does not tell us anything about the quality of the peace, and the quality of the peace can be relevant to the requirements for a stable peace. These requirements have both an empirical basis and a normative basis, as I discuss in Chapter 5.

This study is concerned with measuring peace consolidation in relation to *civil wars* or *internal armed conflicts* (the terms are used interchangeably here), the most prevalent form of armed conflict in the post-Cold War era.[21] Civil wars are armed conflicts between the government of a state and one or more opposition groups within that state or among non-state groups only, although foreign powers may also be involved (as with the armed conflicts in the Democratic Republic of Congo from 1996 or Syria from 2011). There are academic conventions governing the use of the term 'civil wars'—and, by extension, 'peace'—that are concerned with the identity of the actors involved (state, non-state) and the intensity of the conflict (number of battle-related deaths). The primary reason for these conventions is to facilitate comparability across different analyses—quantitative analyses especially. Where precision of terms is required—notably in Chapter 4—the terms are defined precisely. Otherwise, when drawing on cases for illustrative purposes only, this study may depart from these academic conventions (e.g., where the intensity of the conflict falls below the agreed threshold).

Peacebuilding is used in this study to refer to a range of activities by governmental and non-governmental actors to preserve and strengthen the peace following a cessation of major hostilities with the aim of establishing a self-sustaining peace. UN Secretary-General Boutros Boutros-Ghali was one of the first to introduce the term to the diplomatic lexicon in *An Agenda for Peace* (1992), in which he defined 'post-conflict peacebuilding' as 'action to identify and support structures which will tend to strengthen and solidify peace in order to avoid a relapse into conflict'.[22] For the purposes of this study the term will be used broadly to refer to the array of third-party interventions

---

[21] Lotta Themnér and Peter Wallensteen, 'Armed Conflict, 1946–2013', *Journal of Peace Research* 51:4 (2014), 541–54.
[22] 'An Agenda for Peace: Preventive Diplomacy, Peacemaking and Peacekeeping', Report of the Secretary-General pursuant to the statement adopted by the Summit Meeting of the Security Council on 31 January 1992, UN Doc. A/47/277–S/241111, 17 June 1992, II.21.

that contribute to the consolidation of peace, thus blurring the rather bureaucratic and programmatic distinctions that exist, within the UN for instance, between peacekeeping and peacebuilding.[23] In many respects there is no practical distinction between the peacekeeping and peacebuilding phases of an operation; the two can overlap significantly.[24] Indeed, the emerging view within the UN is that peacebuilding is an approach that should inform all forms of UN engagement—before, during, and after violent conflict—an approach captured by the term 'sustaining peace'.[25] Peacebuilding also shares certain functional properties in common with *stabilization*, a term more narrowly associated with the use of military instruments. As this study is concerned primarily with assessing the quality of peace, such distinctions are not always significant for our purposes.

Finally, although the term *post-conflict* is employed here in conjunction with peacebuilding, the term is something of a misnomer.[26] No society is without conflict and even 'peaceful' societies may experience episodes of violence associated with prior conflict.[27] Consider, for instance, the clashes over the use of the Confederate flag in the United States more than 150 years after the formal end of the civil war there.[28] What matters is the degree of violence that a society experiences. The use of the term is conventional, however, and serves usefully to identify the particular phase or phases in the evolution of violent conflict with which this book is concerned. Despite its shortcomings, therefore, the term will be used in reference to the period following the cessation of armed conflict, variably defined.

We turn now to consideration of the term at the centre of this study: peace.

---

[23] For a discussion of how the concept is deployed by the UN, the European Union, donor states, and rising powers, see Charles T. Call and Cedric de Coning, 'Conclusion: Are Rising Powers Breaking the Peacebuilding Mold?' in Call and De Coning (eds), *Rising Powers and Peacebuilding: Breaking the Mold?* (Basingstoke: Palgrave Macmillan, 2017), ch. 10.

[24] Reconciliation efforts, for instance—a key component of peacebuilding—often begin in the peacekeeping phase.

[25] See UN Security Council Resolution 2282 (2016), 27 April 2016.

[26] For a discussion of the post-conflict concept, see Chip Gagnon and Keith Brown (eds), *Post-Conflict Studies: An Interdisciplinary Approach* (Abingdon: Routledge, 2014).

[27] Astri Suhrke and Mats Berdal (eds), *The Peace in Between: Post-War Violence and Peacebuilding* (Abingdon: Routledge, 2012).

[28] 'Charleston Shooting: Confederate Flag at Heart of Growing Political Storm', *Guardian*, 20 June 2015.

1

Conceptualizing Peace

> Peace is too important a goal to be without a firm conceptual basis for both research and positive social action.
>
> Royce Anderson[1]

What is peace? This is not, as it might seem, a pedantic question. Rather, it is a question of fundamental practical importance. Clarifying what is meant by peace is critical to measuring peace consolidation. Without clarity about the characteristics of peace, it is difficult if not impossible to assess progress towards achieving it. However, because peace is an 'essentially contested term', there is no generally agreed definition of it. Is it 'merely' the absence of violent conflict? Does it require the transformation of society and relations within it? Different actors may hold different views about its meaning. The views of government elites may differ from those of rebel leaders or international peacekeepers or donor governments. These various views do not necessarily exist in harmony; indeed, they may be mutually exclusive, which makes the challenge of measuring peace consolidation in the aftermath of civil war all the more difficult.

Notwithstanding this lack of consensus, embedded within every peacebuilding strategy, stated or otherwise, is a particular conception of peace that informs that strategy. Thus, for instance, the Organization for Security and Co-operation in Europe is guided in its approach to what it calls 'post-conflict rehabilitation' by the notion of 'comprehensive security'. As a consequence, the organization takes a broad approach to peacebuilding, extending beyond the cessation of violent conflict to the 'politico-military, economic and environmental,

---

[1] Royce Anderson, 'A Definition of Peace', *Peace and Conflict: Journal of Peace Psychology* 10:2 (2004), 115.

14   *Measuring Peace*

and human aspects of security'.[2] Many national military organizations, by contrast, while mindful of the importance of a comprehensive approach, tend to view peace more narrowly as the condition that obtains with the termination of hostilities.[3] How exactly, and to what extent, conceptions of peace inform, guide, direct, or constrain the practices of peacebuilding organizations requires some elaboration. I examine that question in Chapter 2. In this chapter, I look at the range of meanings that social scientists employ with their use of the term 'peace' (and associated terms) as these meanings allow us to construct a conceptual map on which we can situate the orientations and activities of the various peacebuilding organizations. A conceptual map also allows us to explore the possibility of developing more differentiated understandings of peace that, I will argue, can in turn be used to expand existing practices of assessing progress towards the achievement of a consolidated peace. While some of the conceptions of peace that I examine here, therefore, reflect the way that policy actors actually use the term, many of the conceptions are suggestive of a richer array of possibilities.

MINIMAL AND MAXIMAL CONCEPTIONS
OF PEACE

The most basic conception of peace is reflected in the binary distinction between 'negative' and 'positive' peace. The Norwegian sociologist Johan Galtung introduced this now classical distinction in the inaugural issue of *Journal of Peace Research* in 1964. Negative peace refers to the 'absence of violence, absence of war', whereas positive peace, as Galtung put it then, refers to the 'integration of human society' such that conflict is not eliminated but 'dynamics without recourse to violence is built into the system'.[4] A negative or 'cold'

---

[2] Organization for Security and Co-operation in Europe, *Background Brief: OSCE Activities and Advantages in the Field of Post-Conflict Rehabilitation*, OSCE Doc. SEC. GAL/76/11, 28 April 2011, 1.

[3] Charles T. Call and Elizabeth M. Cousens, 'Ending Wars and Building Peace: International Responses to War-Torn Societies', *International Studies Perspectives* 9:1 (2008), 4.

[4] 'An Editorial', *Journal of Peace Research* 1:1 (1964), 2. Galtung would later revise his definition of positive peace to mean the 'absence of structural violence', likening it

Conceptualizing Peace

peace may exist between adversaries—as with the United States and the Soviet Union during the Cold War—or between formerly warring states—as with Egypt and Israel today—and it may even be a stable peace,[5] although few will question the relatively greater stability of the positive peace that exists between post-war Germany and France, for instance, and the sense of security it affords. Within conflict-affected states it is more commonly thought that a transformation of relations between the parties to the conflict—a positive or 'warm' peace—is required, as might be achieved, for instance, through processes of reconciliation and confidence-building.[6] While a negative peace may suffice in a post-civil war state, a positive peace, it is widely agreed, affords communities greater prospects for preventing the recurrence of violent conflict.[7]

A negative peace is a minimal conception of peace in so far as the absence of armed conflict is the minimum condition for peace. Armed conflict is, quite obviously, antithetical to peace. However, what constitutes armed conflict, what distinguishes peace from war, how armed conflicts differ from other forms of collective violence, and how long a peace must endure to qualify as a peace (as opposed to a mere suspension of hostilities) are not so obvious. These are just a few of the key questions the answers to which are important for clarifying what is meant more precisely by a negative peace.

A minimal conception of peace is reflected in the broad body of statistical studies of armed conflicts. The two most widely utilized data sets in this regard are those developed by the Correlates of War Project (COW) and the Uppsala Conflict Data Program/Peace Research Institute, Oslo (UCDP/PRIO).[8] Both data sets measure

---

to 'social justice'. See below and Galtung, 'Violence, Peace, and Peace Research', *Journal of Peace Research* 6:3 (1974), 167–91.

[5] Kenneth N. Waltz, 'The Stability of a Bipolar World', *Daedalus* 93:3 (1964), 881–909.

[6] John Paul Lederach, *Building Peace: Sustainable Reconciliation in Divided Societies* (Washington, DC: United States Institute of Peace Press, 1997), 20, 82–3.

[7] Johan Galtung, 'Twenty-Five Years of Peace Research: Ten Challenges and Some Responses', *Journal of Peace Research* 22:2 (1985), 141–58.

[8] The Correlates of War Project, http://www.correlatesofwar.org; UCDP/PRIO Armed Conflict Database, https://www.prio.org/Data/Armed-Conflict/UCDP-PRIO. COW and UCDP/PRIO have developed a number of different databases. I am concerned here with the more widely used databases, which contain data on episodes of war/armed conflict, as defined below.

16   *Measuring Peace*

armed conflict; they do not measure peace, not even negative peace. This might seem counterintuitive. After all, peace at a minimum, we have just observed, is the absence of armed conflict. However, while that may be true conceptually, the 'absence of armed conflict' as inferred from these data sets exhibits a range of characteristics, not all of which may be compatible with notions of peace, negative or otherwise. The distinction will become apparent below.

COW and UCDP/PRIO utilize a classification system that is based in part on the status of the parties to the conflict. COW in its present incarnation (it has evolved since it was established in 1963) and UCDP/PRIO employ similar categories of conflict: *interstate wars*, where the parties to the conflict are recognized states; *extra-state* or *extra-systemic wars*, which occur between a state and a non-state group outside the state's territory (e.g., colonial wars); and *intra-state* or *internal conflicts*—now the most prevalent form of armed conflict (see Figure 1.1)—which occur between the government of a state and one or more internal opposition groups. COW has a fourth category—*non-state wars*—which refers to armed conflicts between or among non-state entities. UCDP/PRIO also has a fourth category—*internationalized internal armed conflicts*—which occur between a

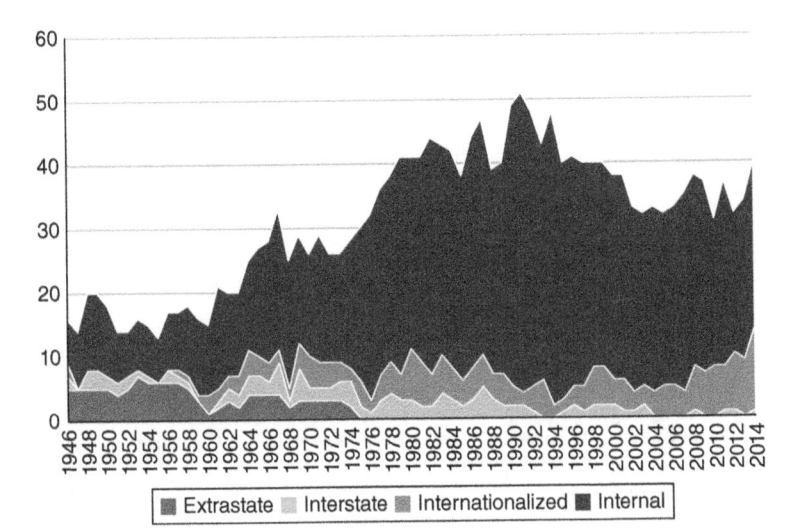

Figure 1.1. Total armed conflict by type, 1946–2014
Source: Pettersson and Wallensteen 2015

Conceptualizing Peace   17

state and one or more armed internal groups with intervention from outside states, as with the Syrian civil war from 2011.[9]

Both COW and UCDP/PRIO employ thresholds of violence to determine whether and when a war or an armed conflict has occurred (the two terms represent different magnitudes of violence). Within the COW typology, an intra-state war must result in 'a minimum of 1,000 battle-related combatant fatalities' in a given year to qualify as a war.[10] UCDP/PRIO uses a lower threshold: 'at least 25 battle-related deaths' (military and civilian) in a given year.[11] The use of a lower threshold in the case of UCDP/PRIO reflects the attempt to capture the broader phenomenon of 'armed conflicts', of which the large majority are now intra-state conflicts. (Another approach would be to establish thresholds based on per capita calculations, which would highlight the *relative* magnitude of violence.)[12] As well as excluding from its principal data set conflicts in which the state is not a party ('non-state conflicts'), UCDP/PRIO also omits 'one-sided violence' that involves the unopposed killing of unarmed civilians, as occurs with massacres.[13] Neither data set, for that matter, takes account of relatively new patterns of violence that elude conventional definitions of collective violence but whose lethality may exceed that of conventional armed conflicts, including gang warfare in Central America, which has had some of the highest homicide rates in the world.[14]

---

[9] Meredith Reid Sarkees and Frank Whelon Wayman, *Resort to War: A Data Guide to Inter-State, Extra-State, Intra-State, and Non-State Wars, 1816–2007* (Washington, DC: CQ Press, 2010), ch. 2; Uppsala Conflict Data Program and International Peace Research Institute, Oslo, 'UCDP/PRIO Armed Conflict Dataset Codebook, Version 4-2014a', http://www.pcr.uu.se/research/ucdp/datasets/ucdp_prio_armed_conflict_dataset. COW's four primary categories are sub-divided further into various sub-categories.

[10] Meredith Reid Sarkees, 'Codebook for the Intra-State Wars v.4.0: Definitions and Variables', 1–2, http://www.correlatesofwar.org/data-sets/COW-war.

[11] Uppsala Conflict Data Program and International Peace Research Institute, Oslo, 'UCDP/PRIO Armed Conflict Dataset Codebook, Version 4-2014a', 1.

[12] See Nicholas Sambanis, 'What Is Civil War? Conceptual and Empirical Complexities of an Operational Definition', *Journal of Conflict Resolution* 48:6 (2004), 821–2.

[13] UCDP/PRIO maintain separate data sets for these conflict dynamics for the period since 1989.

[14] In 2012, Honduras's annual homicide rate of 90.4 per 100,000 population was fifteen times greater than the average global annual homicide rate (6.2 per 100,000 population). Figures from United Nations Office on Drugs and Crime, *UNODC Global Study on Homicide 2013: Trends, Context, Data* (Vienna: UNODC, 2014) 21, 24.

18   *Measuring Peace*

Domestic violence—which has important but poorly understood links to political violence—is not taken into consideration either.[15]

From a conceptual standpoint, then, for the vast number of studies that rely on these data sets, peace is a negative peace—that is, the absence of armed conflict as each data set defines armed conflict. If relations between parties to a conflict are fraught, a state is nonetheless considered to be at peace provided that the threshold in battle-related deaths has not been crossed in a given year. However, the absence of armed conflict may not actually signify the achievement of peace, not even a negative peace, as one might understand that term. In the case of COW, if battle-related fatalities total 999 or less in one year, this qualifies as peace. Indeed, a number of actual armed conflicts persist below the threshold of both data sets in a given year (in Papua New Guinea, Turkey, and Gaza, among other conflict situations). Moreover, the absence of violence, where it does occur, may merely reflect a period of preparation for the resumption of larger-scale fighting. There is also the issue of the distribution of violence (localized versus generalized) in a country and, as indicated earlier, the *relative* impact of conflict intensity (the number of battle deaths relative to the size of the population), neither of which is captured by these data sets. In sum, these data sets imply a notion of negative peace that is both underspecified and undifferentiated, for which reason they do not allow us to measure the quality of the peace precisely on the basis of them.

Negative peace is thought by some scholars to have only limited utility for other reasons. Galtung, with aspirations for transformative social action in mind, rejects this 'narrow concept of peace' on normative grounds because, as he puts it, 'if [killing] were all violence is about, and peace is seen as its negation, then too little is rejected when peace is held up as an ideal. Highly unacceptable social orders would still be compatible with peace.'[16] Negative peace also has its detractors for reasons that are largely empirical: negative peace, as we have seen, is thought to be an insufficient basis for a stable peace.

---

[15] As Lansford and Dodge observe, '[V]iolence in one domain tends to generalize, or spill over, into other domains. For example, war, homicide, assault, combative sports, and severe punishment of criminals jointly characterize cultures of violence.' Jennifer E. Lansford and Kenneth A. Dodge, 'Cultural Norms for Adult Corporal Punishment of Children and Societal Rates of Endorsement and Use of Violence', *Parenting, Science and Practice* 8:3 (2008), 257–70.

[16] Galtung, 'Violence, Peace, and Peace Research', 168.

Conceptualizing Peace 19

The two objections are not unrelated: many (state) authorities who support 'highly unacceptable social orders' lack legitimacy; their regimes are unstable as a result and they are prone, therefore, to use violence domestically to suppress dissent.[17] Others simply view positive peace as a natural complement to negative peace, much the same way as the World Health Organization in its constitution defines health as 'a state of complete physical, social and mental well-being, and not merely the absence of disease or infirmity'.[18]

A positive peace, then, may have intrinsic value and/or instrumental value—i.e., to ensure a stable peace. But what are the characteristics of a positive peace? What state of affairs does it describe? Galtung's conception of a positive peace sets the bar very high: employing an extended notion of violence, latterly understood to occur 'when human beings are being influenced so that their actual somatic and mental realizations are below their potential realizations'.[19] Galtung's positive peace entails the elimination of both direct personal violence and indirect structural violence, i.e., the norms, institutions, attitudes, and other features of societies that inhibit the individuals within them from achieving their full potential.[20] Positive peace conceived in these terms, as we will see in Chapter 2, exceeds what any peacebuilding strategy would normally seek to achieve. Galtung's conception is useful, however, for delineating the upper limits conceptually of major scholarly thinking on this question.

Galtung's conception of positive peace is ambitious, perhaps even utopian,[21] but it has intellectual merit in so far as it represents a Weberian 'ideal type' response to the problem of violence, as Galtung defines it. It is a useful analytical abstraction although it does not (and arguably could not) exist, its chief weakness being that it is

---

[17] Juan J. Linz, *Totalitarian and Authoritarian Regimes* (Boulder, CO: Lynne Rienner, 2000).

[18] Constitution of the World Health Organization, *Basic Documents*, 45th edition, Supplement, October 2006.

[19] Galtung, 'Violence, Peace, and Peace Research', 168.

[20] Galtung would later add 'cultural violence' as a third aspect of violence. See Johan Galtung, 'Cultural Violence', *Journal of Peace Research* 27:3 (1990), 291–305.

[21] 'If peace requires the absence of [all] political and structural violence,' Yan Xuetong observes, 'it has never been experienced in human history, because inequality, destitution, oppression and discrimination have existed in one form or another in all societies through the ages.' Yan Xuetong, 'Defining Peace: Peace vs. Security', *Korean Journal of Defense Analysis* 16:1 (2004), 203.

20   *Measuring Peace*

underspecified and seemingly inexhaustible in its requirements.[22]
Unlike a reduction in the number of battle-related fatalities, it is
difficult if not impossible to determine whether human beings (all?
most?) are living at or beneath their potential. This weakness is shared
by other conceptions of positive peace to the extent that they too
are underspecified and, furthermore, may fail to distinguish what is
necessary from what is desirable with respect to the requirements
for a positive peace. There is a tendency, for instance, to conflate
peacebuilding with state-building, democratization, and other related
activities, resulting in a kind of conceptual inflation when all of the
associated goals and activities are subsumed beneath the single banner
of peacebuilding.[23] Some of the desired outcomes—democratization,
for example—may not be necessary to achieve a stable peace; indeed,
they may even militate *against* a stable peace. In Burundi, for instance,
with the election of the country's first Hutu president following multi-
party democratic elections held in June 1993, the country descended
into violence more deadly than all previous outbreaks of interethnic
violence combined since independence.[24] This conflation of goals is a
problem that manifests itself, too, in peacebuilding practice, as we will
see in Chapter 2.

TOWARDS A MORE DIFFERENTIATED
CONCEPTION OF PEACE

The terms negative peace and positive peace, we have seen, are each
broad in their scope. Within and between these two poles there lies a
wide range of 'post-conflict' conditions. Scholars have sought to
develop conceptually rigorous ways of describing the entire range of

---

[22] Galtung's conception of positive peace has attracted much criticism for this
reason. See, for instance, John Keane, *Reflections on Violence* (London: Verso, 1996)
and C. A. J. Coady, 'The Idea of Violence', *Journal of Applied Philosophy* 3:1 (1986),
3–19.

[23] For a discussion of this problem, see Charles T. Call with Vanessa Wyeth (eds),
Building States to Build Peace (Boulder, CO: Lynne Rienner, 2008), chs 1, 15.

[24] Michael Lund, 'What Kind of Peace Is Being Built? Taking Stock of Post-
Conflict Peacebuilding and Charting Future Directions', mimeo, January 2003, 17.
For a general analysis, see Edward D. Mansfield and Jack Snyder, *Electing to Fight:
Why Emerging Democracies Go to War* (Cambridge, MA: MIT Press, 2007).

Conceptualizing Peace 21

post-conflict conditions and of measuring them. This section examines some of these efforts with an eye towards identifying ways of enriching analysis of the quality of peace.

A number of scholars take issue with the 'sharp categorical distinction between "war" and "peace"', as Paul Richards describes it. He urges analysts to think instead of peace 'in terms of a continuum.'[25] In this vein, Royce Anderson takes as his starting point what he sees as the two dimensions of peace—the absence of violence (negative peace) and the presence of harmonious relations (positive peace) in a conflict-affected society—the extent of which in each case, he suggests, can be plotted along a continuum. These two dimensions, moreover, manifest themselves as objective conditions and subjective perceptions which, Anderson maintains, can both be measured. 'Though peace can be partially measured by objective measures,' he observes, 'comprehensive measures of peace should also include subjective indicators that reflect people's personal evaluations and experience of peace.'[26] Measuring the quality of the peace would require developing indicators for each of Anderson's four components of peace: objective indicators of violence, subjective indicators of violence, objective indicators of harmonious relations, and subjective indicators of harmonious relations. The key prior question, of course, is how narrowly or broadly are the terms 'violence' and 'harmonious relations' to be defined in a given context? I return to this question when I consider ways of operationalizing the conceptions under discussion here. For our purposes now, however, what is important to bear in mind is the utility of a conceptual continuum.

Nadine Ansorg, Felix Haass, and Julia Strasheim also imply a continuum in their conceptualization of peace, which they define as 'a continuous condition characterized by the absence of direct, physical violence between social or political groups, with all relevant actors regarding the non-violent regulation of social and political conflict as the "only game in town"'.[27] Ansorg and colleagues adapt for this

[25] Paul Richards, 'New War: An Ethnographic Approach', in Paul Richards (ed.), *No Peace, No War: An Anthropology of Contemporary Armed Conflicts* (Athens, OH/ Oxford: Ohio University Press/James Currey, 2005), 5.

[26] Anderson, 'A Definition of Peace', 104.

[27] Nadine Ansorg, Felix Haass, and Julia Strasheim, 'Between Two "Peaces"?: Bridging the Gap between Quantitative and Qualitative Conceptualizations in Multi-Method Peace Research', paper presented at the International Studies Association Annual Convention, San Francisco, 3–6 April 2013, 7.

22   *Measuring Peace*

purpose Juan J. Linz and Alfred Stepan's well-known definition of a consolidated democracy: 'a political situation in which, in a phrase, democracy has become "the only game in town"'.[28] In relation to a consolidated peace, this would mean that no significant social or political groups seek resort to violence to achieve their ends and that the majority of the population believes that political differences can be resolved and political change should be sought using only peaceful means.

Ansorg and colleagues propose assessing the strength of peace at three different levels of society: micro, meso, and macro. The micro level concerns the attitudes and behaviour of the general population, and the extent to which political violence is accepted/rejected or employed by members of the general public. The meso level concerns the attitudes and behaviour of groups and whether they support or engage in the use of violence to challenge the government or to attack other groups. A key indicator, in this regard, is the continued existence of armed opposition movements.[29] The macro level is concerned with the views and behaviour of political and bureaucratic elites and whether they sanction the use of force in settling political differences. In each case one can imagine a range of possible dispositions/behaviours that could generate a value for the quality of the peace for the state as a whole. The assessment could be conducted on a regional basis to capture variations within a state, which is extremely important given that intra-state conflicts are often highly localized (the post-Mobutu conflicts in the Democratic Republic of Congo are a prime example).[30] This is not a dynamic analytical framework, however, and therefore it would not by itself convey a sense of whether the situation is improving, deteriorating, or remaining the same.

A more fully developed 'peace scale' is suggested by James Klein, Gary Goertz, and Paul Diehl (see Table 1.1).[31] Although developed

---

[28] Juan J. Linz and Alfred Stepan, *Problems of Democratic Transition and Consolidation: Southern Europe, South America and Post-Communist Europe* (Baltimore, MD: Johns Hopkins University Press, 1996), 5.

[29] Adrian Florea, 'Where Do We Go from Here? Conceptual, Theoretical, and Methodological Gaps in the Large-N Civil War Research Program', *International Studies Review* 14:1 (2012), 82, fn. 19.

[30] Nils Weidmann, Jan Ketil Rød, and Lars-Erik Cederman, 'Representing Ethnic Groups in Space: A New Dataset', *Journal of Peace Research* 47:4 (2010), 491–9.

[31] James P. Klein, Gary Goertz, and Paul F. Diehl, 'The Peace Scale: Conceptualizing and Operationalizing Non-Rivalry and Peace', *Conflict Management and Peace Science* 25 (2008), 67–80.

Table 1.1. Core features of assessing the state of peace or rivalry between countries

Indicator	Peace scale indicators				
	Rivalry	Negative peace		Positive peace	
	Rivalry	Low-level conflict	Negative peace	Positive peace	Pluralistic security community
	1.00	0.75	0.50	0.25	0.00
	Conflict				
1. War plans	Present	Present	Absent	Absent	Joint war planning
2. Conflicts	Frequent MIDs; variety of hostility levels	Isolated MIDs; Thompson rivalries; ICB crises	Absent	No plausible counterfactual war scenarios	No plausible counterfactual war scenarios
	Communication and issues				
3. Main issues in conflict	Unresolved	Unresolved	Mitigated; some resolved; some low salience	Resolved	Resolved
4. Communication	Absent	Absent	Intergovernmental	Intergovernmental and highly developed transnational ties	Institutionalized mechanisms
	Agreements, institutions, and diplomacy				
5. Diplomacy	No recognition; diplomatic hostility	No recognition or diplomatic hostility	Diplomatic recognition; statements suggesting conflict	Diplomatic relations	Diplomatic coordination
6. Area/level/number Agreement	None	None	Peace negotiations and/or agreements	Nascent functional agreements; nascent integration	Extensive institutionalized functional agreements

Note: MIDs = militarized interstate disputes; ICB = international conflict behaviour
Source: Klein et al. 2008

24   *Measuring Peace*

for the purpose of assessing the quality of relations between states, it
has potentially useful application—in modified form—to intra-state
conflict dynamics. The authors construct a scale with rivalry at one
end (1.0), a pluralistic security community at the other end (0.0), and
low-level conflict (0.75), negative peace (0.5), and positive peace
(0.25) between the two poles. Corresponding to each of the five points
on the scale are a series of descriptive indicators that provide further
specification. While the terms are not all appropriate for intra-state
conflict, the table exemplifies the use of the notion of gradations or
degrees of peace.

The final conceptualization of peace considered here is reflected in
Kristine Höglund and Mimmi Söderberg Kovacs's notion of the
'Peace Triangle'.[32] Höglund and Söderberg Kovacs are interested in
the nature of the peace that prevails following a negotiated settlement
and observe, consistent with the foregoing discussion, that the character and quality of the peace can vary significantly. To capture this
variation, they distinguish post-settlement peaceful societies with
regard to three categories: issues, behaviour, and attitudes. The first
category, issues, is concerned with whether or not all (major) conflict
issues have been resolved. Many peace agreements defer resolution of
particularly difficult issues until a later date in the hope that the
experience of peace will build confidence between the parties and
thus facilitate resolution. Höglund and Söderberg Kovacs refer to this
as an *unresolved peace*. The Oslo Peace Accord (1993) between
Palestinians and Israelis, for instance, left many issues unresolved,
notably the status of Jerusalem and Palestinian refugee returns.
A peace may also be a *restored peace*, with, for example, the removal
of a tyrant or the defeat of rebel forces, or a peace may be a *contested
peace* when, for instance, victorious armed opposition groups turn
against one another.

The second leg of the Peace Triangle concerns conflict behaviour
after a peace agreement—whether one or another of the warring
parties continues to resort to violence (*partial peace*); whether some
but not all of a country is pacified with the signing of a peace
agreement (*regional peace*); and whether insecurity prevails as a result
of widespread criminal behaviour that perhaps stems directly from

[32] Kristine Höglund and Mimmi Söderberg Kovacs, 'Beyond the Absence of War:
The Diversity of Peace in Post-Settlement Societies', *Review of International Studies*
36:2 (2010), 367–90.

Conceptualizing Peace

the war itself or feeds on adverse post-war conditions such as high unemployment (*insecure peace*).

The final leg of the Peace Triangle concerns the relative presence or absence of conflict attitudes in the period following a peace agreement. A conflict-affected population may remain seriously divided after an agreement (*polarized peace*), as was the case with Republicans and Loyalists in Northern Ireland following the Good Friday Agreement in 1998. There may also be a sense among groups that peace has been established on an unfair basis, perhaps as the result of the failure to prosecute war criminals (*unjust peace*). Or peace may be established on the basis of a strong man coming to power who may not enjoy broad popular legitimacy (*fearful peace*).

What these various initiatives all have in common is a recognition of the highly heterogeneous character of peace. They demonstrate that peace is a much more complex concept than the simple binary distinction between negative and positive peace would suggest. They also point to, if not explicitly state, the value of being able to conceptualize peace in such a way as to allow us to make distinctions on the basis of both different kinds of peace and different degrees of peace. The question is whether these conceptualizations can also be operationalized and employed to help ascertain whether a particular peace is a self-sustaining peace. We examine this question in Chapter 5.

OTHER CONCEPTS

Scholars (and practitioners) employ other concepts that are closely related to peace. These terms can be suggestive of ways of further conceptualizing peace; they also provide perspective on the limitations, apparent or real, of the term. For both reasons they have potential value in relation to the practice of assessing the quality of post-conflict peace.

Klein, Goertz, and Diehl do not propose new terminology so much as take issue with the notion of a 'stable peace', which is a key concept for the purposes of this study. They point out that stability can be achieved at all points on the peace continuum, from the coldest negative peace to the warmest positive peace. Stability is simply an equilibrium in relations, which can be of fairly poor quality. 'Stability,' they conclude

26   *Measuring Peace*

for this reason, 'is, of course, a very important theoretical and empirical
question, but not one that we should build into the concept of peace.'[33]

It is true that if stability in itself were the primary objective, then
the quality of peace in other respects might suffer. However, from an
analytical standpoint, surely it must be valuable to know how stable a
peace may be, whatever its character in other respects. (This may be
their point about stability being 'a very important theoretical and
empirical question'.) To seek to ascertain if a peace is stable is not to
seek to establish a stable peace. On the other hand, a stable peace may
indeed be the objective, in which case whether it is a cold peace will
only be a function of its stability if stability can only be achieved at a
'low level' of peace. There may be circumstances in which that is the
case—for instance, if the costs associated with instability are high and
the pursuit of a warmer peace entails risks of instability that are
thought to be too great. Kosovo has arguably been such a case,
where Serbs and Albanians have been so deeply divided that to date
it has not been possible to pursue anything more than a cold (but
stable) peace in some parts of the territory. Once a modicum of
stability has been achieved, moreover, it may become possible to
enhance the peace over time if the political environment improves.[34]
Stability thus has its value and it does not necessarily debase the
quality of the peace.

Security and peace are closely related but distinct concepts. Security exists in the absence of dangers and threats. It is a broader concept,
in relation to peace, that has broader functional applicability. One can
talk about 'physical security', 'economic security', 'environmental
security', 'food security', etc., as well as, more generally, 'human
security'. The idea of human security represents a shift from the
state as the principal referent of security to the individuals within it.
As with the notion of positive peace, human security is vulnerable to
'conceptual overstretch'—the extension of the concept so that it
applies to almost every conceivable danger or threat.[35]

---

[33] Klein, Goertz, and Diehl, 'The Peace Scale', 69.
[34] This thinking is reflected in the UK Foreign and Commonwealth Office, Ministry
of Defence, and Department for International Development's *Building Stability Framework*, 2016, which calls for building 'fair power structures that broaden inclusion,
accountability and transparency *over time*' (emphasis added).
[35] S. Neil MacFarlane and Yuen Foong Khong, *Human Security and the United
Nations: A Critical History* (Bloomington, IN: Indiana University Press, 2006), 228.

Conceptualizing Peace

It is possible to achieve peace, even a stable peace, without security.[36] Many people across the globe live in peaceful conditions but face daily threats to their security—ethnic minorities, for instance, who live in constant fear of discrimination and abuse, or communities that fear the predations of corrupt police. Peace and security are complementary: threats to security, if severe enough, may threaten the peace, while it can be argued that an expansive notion of peace can and should embrace security. Indeed, peace without security, one might say, is really just 'negative peace' by another name. Nonetheless, it is important to bear the distinction in mind if only to ascertain, among other things, if the peace that has been established has taken security sufficiently into consideration.

Another notion that is used with increasing frequency by scholars and practitioners in relation to fragile and conflict-affected states is 'resilience'.[37] Resilience is the capacity to withstand internal or external shocks or pressures. Resilient societies are able to absorb these blows and to recover from them. It does not mean that they remain unaltered as a consequence of the shocks and pressures; to the contrary, they may need to adapt in order to cope with the challenges.[38] But adaptive capacity is a characteristic feature of a resilient society.[39]

The notion of resilience represents a different approach to coping with the threats of violent conflict. It places the emphasis on characteristics of a society—its norms, practices, culture, and institutions—that serve to insulate it from threats rather than on an extrinsic condition (stable peace) and what is required to achieve and maintain it. One implication of this difference is that thinking about resilience-building sits comfortably within a developmental framework, which tends to have a longer-term perspective. As the World Bank observed in its 2011 *World Development Report*,

---

[36] Yan, 'Defining Peace', 207.

[37] Ken Menkhaus, 'Making Sense of Resilience in Peacebuilding Contexts: Approaches, Applications, Implications', *Geneva Peacebuilding Platform Paper 6* (2013), http://www.gpplatform.ch/publications.

[38] The US Agency for International Development (USAID) defines resilience as 'the ability of people, households, communities, countries and systems to mitigate, adapt to and recover from shocks and stresses in a manner that reduces chronic vulnerability and facilitates inclusive growth'. USAID, 'Building Resilience to Recurrent Crisis: USAID Policy and Program Guidance', December 2012.

[39] David Chandler, 'Resilience and Human Security: The Post-Interventionist Paradigm', *Security Dialogue* 43:3 (2012), 217.

28   *Measuring Peace*

Creating the legitimate institutions that can prevent repeated violence is, in plain language, slow. It takes a generation. Even the fastest-transforming countries have taken between 15 and 30 years to raise their institutional performance from that of a fragile state today—Haiti, say—to that of a functioning institutionalized state, such as Ghana.[40]

By contrast, the time horizon for peace-building has tended to be more short term, as many critics have pointed out.[41] A corollary is that resilience-building is a process that external parties can at best support but cannot engineer, as Ken Menkhaus observes:

> The qualities that make up resilience are deeply embedded, not virtues that can be quickly transferred in a workshop. They involve dense patterns of trust networks, hybrid coalitions forged across a wide range of actors, shared narratives, common interests, multiple lines of communication, good leadership, and a commitment by local leaders to take risks for peace—whether this includes negotiating with or confronting potential armed spoilers.[42]

David Chandler suggests further that in so far as resilience entails capacity- and capability-building, it also implies local empowerment: 'The resilience paradigm clearly puts the agency of those most in need of assistance at the centre, stressing a programme of empowerment and capacity-building.'[43] Conceptions of security and peace, by contrast, are consistent with externally furnished protection, although arguably the notion of a *self-sustaining* peace also implies local empowerment. The critical question, often overlooked by practitioners, is who exactly is to be empowered and empowered to do what? We examine this question in Chapter 2.

Stability, security, resilience—these are all concepts that are distinct from peace but that relate to and can enrich our understanding of peace as both an analytical tool and a policy objective. Like peace, they too have their limitations: the terms are imprecise and there are no agreed conventions governing their use. However, they offer the possibility of refining further a concept so obviously fundamental to the enterprise of peacebuilding. There are other ways of

---

[40] World Bank, *World Development Report 2011: Conflict, Security and Development* (Washington, DC: World Bank, 2011), 10.

[41] See, for instance, Ashraf Ghani and Clare Lockhart, *Fixing Failed States* (New York: Oxford University Press, 2009), 5, 108.

[42] Menkhaus, 'Making Sense of Resilience in Peacebuilding Contexts', 7.

[43] Chandler, 'Resilience and Human Security', 216.

Conceptualizing Peace 29

conceptualizing peace.[44] The foregoing discussion by no means exhausts all possibilities but it does reflect the major associated conceptions among them.

* * *

The concept of peace, and the empirical reality of peace, are far more varied and heterogeneous than the scholarly literature often acknowledges. Peace understood as simply the absence of war or armed conflict certainly has utility for the purpose of analysing patterns of violent conflict across space and time. However, it is of very limited value for the purpose of understanding the nature or the quality of a given peace and thus for devising appropriate peacebuilding strategies and for assessing how consolidated a peace may be. This chapter has examined various ways in which scholars have sought to conceptualize peace in an effort to provide, as Höglund and Söderberg Kovacs put it, 'a more fine-grained picture of peace in societies that have come out of civil wars'.[45] Chapter 2 will explore how these and other conceptual distinctions map onto actual peacebuilding practice.

---

[44] See Oliver Richmond's wide-ranging discussion of peace in Oliver P. Richmond, *The Transformation of Peace* (Basingstoke: Palgrave Macmillan, 2005).
[45] Höglund and Söderberg Kovacs, 'Beyond the Absence of War', 389.

2

From Conception to Practice

> Very frequently the 'world images' that have been created by ideas have, like switchmen, determined the tracks along which action has been pushed.
>
> Max Weber[1]

Chapter 1 explored various ways in which scholars conceptualize peace. These conceptions of peace serve several purposes: to describe the range and richness of important empirical phenomena; to facilitate analysis of patterns of armed conflicts and their aftermath; and to articulate goals and aspirations that individuals, governments, and organizations worldwide are engaged in efforts to achieve.

This chapter examines how these conceptions map onto and, to a certain extent, contribute to the actual practice of peacebuilding, with reference to the principal multilateral peacebuilding actors, notably the United Nations (UN); the Organization for Security and Co-operation in Europe (OSCE), the North Atlantic Treaty Organization (NATO) and other regional security organizations; the World Bank; and leading non-governmental organizations (NGOs). What are the primary features of these organizations' approaches to peacebuilding? How do they differ in their understandings of the characteristics of, and requirements for, a consolidated peace? What explains these differences and what are the implications of these differences for the formulation of coherent approaches to peacebuilding?

Two claims about the salience of conceptions underpin the analysis of this chapter. The first claim regards the relationship between ideas

---

[1] Max Weber, 'Social Psychology of the World's Religions' in H. H. Gerth and C. Wright Mills (eds), *From Max Weber: Essays in Sociology* (New York: Oxford University Press, 1958 [1913]), 280.

From Conception to Practice 31

and practice.[2] In Chapter 1 I asserted that conceptions of peace inform, guide, direct, and may even constrain the practices of peacebuilding organizations. What do I mean by that more precisely? I am not suggesting that conceptions of peace are determinative of particular outcomes. That would be too strong a claim. Rather, I would argue that, like Weber's switchmen directing action onto particular tracks, conceptions of peace serve at the very least to orient practitioners broadly, with some practitioners tending as a consequence to view peace largely, but not necessarily exclusively, in terms of security, others in terms of development, and so forth.

More than just orient practitioners, however, different conceptions of peace may also prescribe or proscribe particular practices. The relationship between conceptions and practice in this respect has been evident, for instance, in the humanitarian work of the International Committee of the Red Cross and Médecins sans Frontières (Doctors without Borders), each of which has at times been guided by different conceptions of neutrality in their provision of humanitarian assistance—the latter sometimes rejecting a strict neutrality that might inhibit public condemnation of the conduct of one party or another to a conflict.[3] These differences have both limited and facilitated the work of these organizations in various ways. Conceptions of peace can function in a similar fashion, as we will see, guiding peacebuilders in the adoption of particular measures that they may deem necessary to achieve peace and to ensure its continuation.

Conceptions of peace may be embedded within organizations, providing one of the core components of an organizational culture that shapes the collective understandings of its staff.[4] Functional differentiation, independent of culture, may also affect how an organization's personnel perceive the challenges of building a stable peace: the division of labour among peacebuilding organizations means that different organizations concentrate on different aspects of peacebuilding, although there is also considerable overlap. Practitioners

---

[2] On the relationship between ideas and practice generally, see Albert S. Yee, 'The Causal Effects of Ideas on Policies', *International Organization* 50:1 (1996), 69–108.

[3] Rony Brauman, 'Médecins sans Frontières and the ICRC: Matters of Principle', *International Review of the Red Cross* 94:888 (Winter 2012), 1523–35.

[4] The literature on organizational culture is extensive. For a useful discussion of organizational culture (in relation to military doctrine), see Elizabeth Kier, *Imagining War: French and British Military Doctrine between the Wars* (Princeton, NJ: Princeton University Press, 1997).

may also have the particular outlook on peace and peacebuilding that they do because of their professional formation, whether as diplomats, development economists, soldiers, etc. Ideas (about peace) and practices (of peacebuilding), in this regard, may be mutually constitutive: practitioners may approach peacebuilding from the perspective of their professional background which a particular organizational culture or functional orientation may reinforce.

The second claim, closely related to the first, concerns the relationship between norms and outcomes. Conceptions comprise norms and these norms establish the parameters of 'appropriate' peacebuilding practice. Peacebuilding, as with much other international behaviour, is a rule-governed sphere of activity whose agents are influenced in their conduct by notions, sometimes only tacit, of what they regard as true, reasonable, and right.[5] Peacebuilding organizations generally feel constrained to operate in conformity with these norms rather than to violate them, although for various reasons—for instance, budgetary considerations or considerations of national interest—an organization may not always respect these norms.

An example may help to clarify this point. Broadly speaking, the leading peacebuilding organizations seek to promote liberal democratic norms in conflict-affected states—as evidenced by their mandated support for free and fair elections, independent media, the monitoring of human rights, and strengthening the rule of law, among other liberal practices and institutions.[6] While peacebuilding actors arguably bear some responsibility for reinforcing the autocratic behaviour of some national elites—Joseph Kabila in the Democratic Republic of Congo and Salva Kiir in South Sudan, for instance[7]—it is inconceivable that a leading peacebuilding organization would *overtly* promote the establishment of autocratic forms and practices of governance in conflict-affected territories. Autocracy is not the preferred outcome; liberal democratic norms are considered to be appropriate. The irony is that autocracy may actually make a greater contribution

---

[5] For a discussion of the 'logic of appropriateness', see James G. March and Johan P. Olsen, 'The Logic of Appropriateness', in Robert E. Goodin, Michael Moran, and Martin Rein (eds), *The Oxford Handbook of Public Policy* (Oxford: Oxford University Press, 2008), 689–708.

[6] See, for instance, UN Security Council Resolution 1996 outlining the nature of UN peacebuilding engagement in South Sudan. UN Doc. S/RES/1996 (2011), 8 July 2011.

[7] Richard Gowan, 'Happy Birthday UN: The Peacekeeping Quagmire', *Georgetown Journal of International Affairs* 16:2 (12 August 2015).

From Conception to Practice   33

to the consolidation of peace because of the polarizing effects that democratization can have on states—especially formerly non-democratic states—emerging from violent conflict.[8] Based on their study of sixty-eight post-conflict episodes, Paul Collier, Anke Hoeffler, and Måns Söderbom found: 'Severe autocracy appears to be highly successful in maintaining the post-conflict peace . . . If the polity is highly autocratic, the risk [of reversion to conflict] is only 24.6 percent; whereas if it is not highly autocratic the risk more than doubles to 62 percent'.[9] The authors recognize that liberal democratization may be an intrinsically desirable objective but, they caution, it should not be thought that democratization will increase the durability of a post-conflict peace.

It is hard to imagine the UN or any other peacebuilding organization embracing this inconvenient truth to the extent of actually advocating autocracy over liberal democracy in their peacebuilding strategies (although there is growing recognition among some practitioners that it may at times be necessary to strike bargains with unsavoury local elites in an effort to reduce violence).[10] If liberal democracy is considered to be the more appropriate form of governance for countries emerging from violent conflict, even in the face of evidence to the contrary, what is the basis of this persistent preference? The preference arguably reflects the view that states with well-entrenched liberal democratic norms and practices—notably norms of tolerance and peaceful dispute-resolution mechanisms—are less likely to succumb to civil war.[11] Roland Paris goes further and

---

[8] Roland Paris, *At War's End: Building Peace after Civil Conflict* (Cambridge: Cambridge University Press, 2004), ch. 9; Edward D. Mansfield and Jack Snyder, *Electing to Fight: Why Emerging Democracies Go to War* (Cambridge, MA: MIT Press, 2007).

[9] Paul Collier, Anke Hoeffler, and Måns Söderbom, 'Post-Conflict Risks', *Journal of Peace Research* 45:4 (2008), 470. Similarly, in their study of the period 1816–1992, Hegre et al. conclude that semi-democracies (regimes intermediate between a democracy and an autocracy) are more conflict-prone than autocracies. Håvard Hegre, Tanja Ellingsen, Scott Gates, and Nils Petter Gleditsch, 'Toward a Democratic Civil Peace? Democracy, Political Change, and Civil War, 1816–1992', *American Political Science Review* 95:1 (2001), 17–33.

[10] See, for instance, the 2018 'synthesis paper' produced by the UK Stabilisation Unit's Elite Bargains and Political Deals Project: http://www.sclr.stabilisationunit.gov.uk/publications/elite-bargains-and-political-deals.

[11] Former UN Secretary-General Kofi Annan's remarks in 2000 are representative of this view: 'There are many good reasons for promoting democracy, not the least . . . is that when sustained over time, it is a highly effective means of preventing

34 *Measuring Peace*

suggests that peacebuilding actors are constrained by the 'international normative environment', which exerts a significant influence on peace operation design. In his article 'Peacekeeping and the Constraints of Global Culture', he argues that peacekeeping agencies and their member states are 'predisposed to develop and implement strategies that conform with [prevailing] norms...and are disinclined to pursue strategies that deviate from these norms'.[12] The prevailing norms, in Paris's view, are those generated by the 'global culture', which comprises 'the formal and informal rules of international social life'. Liberal democratic values, among other broadly accepted norms, he observes further, are constitutive of this global culture. Conceptions of peace, similarly, are expressive of norms that establish appropriate standards of political behaviour and institutional development.

Conceptions of peace, therefore, while not a causal factor, are not epiphenomenal either, as some scholars would maintain.[13] Conceptions of peace provide a lens through which peacebuilding actors perceive the characteristics of peace and the requirements for peace maintenance. Different conceptions of peace have different implications for devising strategies of peacebuilding and peace consolidation. What it takes to achieve a minimal peace is very different from what is required to achieve a maximal peace. Conceptions of peace thus help to shape peacebuilding agendas, even if only implicitly.

CONCEPTIONS IN PRACTICE

So how then do peacebuilding organizations conceptualize peace? This section examines the conceptions that inform the work of

conflict, both within and between states'. Kofi Annan's closing remarks to the conference 'Towards a Community of Democracies', Warsaw, 28 June 2000, UN Press Release SG/SM/7467.

[12] Roland Paris, 'Peacekeeping and the Constraints of Global Culture', *European Journal of International Relations* 9:3 (2003), 443. See also Séverine Autesserre, 'Construire la paix: conceptions collectives de son établissement, de son maintien et de sa consolidation', *Critique Internationale* 51 (2011), 153–67.

[13] For a discussion of how the dominant schools of International Relations view ideational factors, see Martin Hollis and Steven Smith, *Explaining and Understanding International Relations* (Oxford: Clarendon Press, 1991), 85, 184, 206.

From Conception to Practice   35

leading peacebuilding organizations. For the purposes of this analysis I have relied on a variety of sources, including organizational guidance notes, planning and implementation documents, evaluation reports, and interviews with peacebuilding practitioners. A number of caveats should be borne in mind. First, it is difficult to generalize about the thinking and practice within organizations, especially in the case of large organizations—e.g., the US government or the UN—with several departments or agencies engaged in different aspects of peacebuilding. The departments or agencies may differ from one another—albeit not radically—in their understandings of and approaches to peacebuilding. Also, ideas and practices with respect to peace and peacebuilding evolve. It can be difficult, therefore, to generalize about the work of an organization over a significant period of time. Finally, it is not always possible to isolate peacebuilding from other conjoined activities—state-building, for instance—whose goals can easily be conflated with those of peacebuilding. Thus, for example, while the international financial institutions are not engaged overtly in peacebuilding, their work—and the conceptions (of effective state institutions, for instance) that underpin that work—are consequential for the nature of the peace that is built.

With these caveats in mind, the first thing to note is that there is a general convergence of thinking (normative as well as conceptual) among the major international peacebuilding actors. Notwithstanding differences—and there are some significant ones—these actors subscribe to a common vision of what peace is (or ought to be) and what is required to sustain it. Broadly speaking, all major peacebuilding organizations share a comprehensive (i.e., positive) conception of peace, even if in their own approaches they may focus on particular aspects of peacebuilding to the exclusion of others. To varying degrees, moreover, all major organizations work in support of the twin pillars—political and economic—of the dominant liberal peacebuilding paradigm and the promotion, as a result, of inclusive political institutions, human rights, the rule of law, and a market economy.[14] In practice, many of the differences among organizations thus tend to be either matters of emphasis—often reflecting functional differentiation—or differences with regard to the modalities of

---

[14] For further discussion of the liberal peacebuilding paradigm, see Paris, *At War's End*, ch. 2, and Oliver P. Richmond, *The Transformation of Peace* (Basingstoke: Palgrave Macmillan, 2005), ch. 5.

implementation (e.g., timing, sequencing, evaluating) of peacebuilding measures, or casualties of political expedience (i.e., the seeming need to support autocratic 'strongmen' already in place).[15]

The United Nations

The largest multilateral peacebuilding actor, in terms of operations and, to a great extent, intellectual leadership, is the UN. The UN has had lead responsibility for peacebuilding in scores of conflict-affected countries since the end of the Cold War. Numerous UN departments, programmes, funds, and specialized agencies contribute to peacebuilding, chief among them the Department of Political Affairs (DPA); the Department of Peacekeeping Operations; the Peacebuilding Commission, the Peacebuilding Support Office, and the Peacebuilding Fund; the UN Development Programme; the UN Children's Fund (UNICEF); and UN Women.

Although the achievement of a consolidated or self-sustaining peace is the stated goal of many UN bodies engaged in peacebuilding, there is no standard or agreed definition of the term among these bodies. Indeed the term, or similar terms, is often used simply to signify the intended outcome of the processes and actions undertaken as part of peacebuilding.[16] Of course, many key terms in the UN Charter elude agreed definition, such as 'threats to the peace' and the 'right of self-defence'.[17] This permits a degree of conceptual flexibility that can be useful, even necessary, but it can also handicap efforts to render terms analytically operational for the purposes of measurement, evaluation, and comparative assessment. Whether or not peace has been achieved, after all, depends in large part on what one means by the term 'peace'.

---

[15] I exclude from consideration primarily state-led peacebuilding programmes, many of which have been illiberal in nature, as with Angola, Rwanda, and Eritrea. See Will Jones, Ricardo Soares de Oliveira, and Harry Verhoeven, 'Africa's Illiberal State-Builders', Working Paper 89 (2013), Refugee Studies Centre, University of Oxford.

[16] United Nations, 'Measuring Peace Consolidation and Supporting Transition', interagency briefing paper prepared for the United Nations Peacebuilding Commission (March 2008), 4.

[17] It was not until 1974 that agreement was reached on a definition of 'aggression'. United Nations General Assembly Resolution 3314, UN Doc. A/RES/3314(XXIX), 14 December 1974.

From Conception to Practice 37

Notwithstanding this lack of definitional precision, there is general recognition across the UN system that peace, if it is to be sustainable, needs to be a 'comprehensive peace'—one that achieves the following three broad and interrelated objectives: the consolidation of security (internal and external); the establishment of effective and inclusive political institutions, norms, and practices; and the fostering of conditions for economic and social rehabilitation, transformation, and development.[18] In support of these broad objectives—each of which, of course, requires further specification—the UN undertakes a wide range of peacebuilding activities. The first objective, the provision of security, will often entail the deployment of peacekeepers and/or military observers; security-sector reform, including the creation of an impartial police force; disarmament, demobilization, and the reintegration of ex-combatants; judicial and penal reform; and mine clearance. The second objective involves the (re-)creation and strengthening of political institutions, political parties, and other participatory mechanisms; capacity-building for government and civil society; regulation of the media; electoral assistance; efforts to curb corruption; and human rights assistance. The third objective is achieved with economic and social development; the return and reintegration of refugees and displaced persons; national reconciliation; the provision of social services; the generation of sustainable sources of livelihood, especially for youth and demobilized soldiers; and judicial and non-judicial measures to redress human rights abuses. This list is indicative, by no means exhaustive, of UN peacebuilding activities, but it serves to demonstrate the comprehensive nature of the peace which the UN often strives to achieve.

While there is general agreement within the UN on the need for a comprehensive approach to peacebuilding, there is not necessarily a consensus among UN practitioners or member states—any more than there is among scholars—as to the specific requirements for achieving a consolidated peace. The use of force in support of UN mandates; the choice of electoral regimes; the limits of inclusivity, the nature of poverty-reduction measures—these are just a few of the

---

[18] As reflected in the landmark documents 'No Exit without Strategy: Security Council Decision-Making and the Closure or Transition of United Nations Peacekeeping Operations', Report of the UN Secretary-General, UN Doc. S/2001/394, 20 April 2001; and, more recently, 'Peacebuilding and Sustaining Peace', Report of the Secretary-General, UN Doc. A/72/707–S/2018/43, 18 January 2018.

many wide-ranging issues that have been subjects of discussion and debate within the organization about how best to achieve a consolidated peace.[19] They demonstrate that there are genuine differences of opinion among the various UN entities and member states as to the most effective and appropriate forms of external engagement. Relatedly, there is also a lack of agreement about the critical factors for measuring peace consolidation.[20]

Alongside these and other 'intellectual' divides, bureaucratic competition among agencies can sometimes explain the differences that arise within the organization with respect to strategic vision. As Oliver Westerwinter observes:

> [S]trategic actors compete with one another for influence over the specific content of peacebuilding strategies ... [G]overnments of developing and developed countries, international financial institutions, UN agencies, and civil society organizations typically disagree about what particular strategic goals to select, the means to accomplish them, and the sequence in which they should be pursued, because *different solutions favor the interests of some actors at the expense of others.*[21]

Evidence in support of this observation can be found in the UN peacebuilding efforts in South Sudan. In July 2011 the UN Security Council authorized the establishment of the UN Mission in South Sudan for the purpose of consolidating the peace in the newly independent state of South Sudan.[22] In view of the threats to security (largely internal) that South Sudan faced, the UN Security Council took the decision to establish an integrated mission with a significant military component. The DPA, historically the lead UN agency on peacebuilding, had argued for consideration of an option that would have resulted instead in the establishment of a special political mission, without a significant military component, which it would have led. The merits of such a mission, the DPA argued, was that it would

---

[19] Interviews with UN officials, New York, October–November 2014.

[20] United Nations Peacebuilding Community of Practice (CoP) e-discussion, 'Measuring Peace Consolidation and Supporting Transition: Summary of Responses', 15 May 2008, https://groups.undp.org/read/messages?id=245888.

[21] Oliver Westerwinter, 'The Informal Powers of International Secretariats: Informal Governance and Brokerage in United Nations Post-Conflict Peacebuilding', paper presented at Nuffield College, Oxford, 6 December 2013, 5 (emphasis added).

[22] UN Security Council Resolution 1996 (2011) authorized the deployment and the establishment of the United Nations Mission in South Sudan, UN Doc. S/RES/1996 (2011), 8 July 2011.

From Conception to Practice   39

not generate 'unfair expectations on protection of civilians that arise
with blue helmets on the ground'.[23] But was it merit or bureaucratic
self-interest that inspired the DPA to argue the case for an option that
it would lead? '[O]rganizations are likely to favor those strategies and
definitions that will most clearly advantage their bureaucratic interests,' Michael Barnett and associates observe in their study of peacebuilding.[24] As it happens, the Security Council chose the more robust
option, authorizing the deployment of up to 7,000 troops, which, in
fact, would prove to be inadequate to deal with the security crisis that
would erupt in South Sudan two years later.[25]

The *point de capiton*—literally the point where the fabric of a sofa
is gathered and anchored by an upholstery button—is used by the
French philosopher Jacques Lacan to describe how in a world of
multiple and contested meanings 'the signifier stops the otherwise
endless movement (*glissement*) of signification' to achieve the illusion
of a fixed meaning.[26] It is a useful term to describe how disparate
activities by the UN, all in support of a common objective (peacebuilding) but sometimes pulling in slightly different directions, are
held together. The *point de capiton* is peace—a consolidated or a self-
sustaining peace.

Organization for Security and Co-operation in Europe

Of the various regional security organizations, the OSCE is the largest
in terms of membership (fifty-seven participating states). It is
engaged in conflict management in relation to all aspects of the
conflict cycle, from conflict prevention to peace consolidation.

The OSCE is concerned primarily with the enhancement of security
within and between its participating states. Its conception of security
is a comprehensive one, reflected in the three 'baskets' or dimensions
of security that were agreed among the original signatories to the

---

[23] 'South Sudan: Planning for Post-CPA UN Presence: Submission to the Policy
Committee', undated memorandum (2011).
[24] Michael Barnett, Hunjoon Kim, Madalene O'Donnell, and Laura Sitea,
'Peacebuilding: What Is in a Name?' *Global Governance* 13:1 (2007), 48.
[25] For an account of the security crisis in South Sudan that erupted in December
2013, see United Nations Security Council, 'Report of the Secretary-General on South
Sudan', 6 March 2014, UN Doc. S/2014/158.
[26] Jacques Lacan, *Écrits* (Paris: Éditions du Seuil, 1966), 805.

Helsinki Final Act in 1975 and have been reaffirmed subsequently: the political-military dimension concerned with basic principles governing relations among the states (sovereignty, inviolability of frontiers, non-intervention, etc.) and confidence- and security-building measures; the economic-environmental dimension concerned with economic cooperation and environmental protection; and the human dimension concerned with human rights and democratic processes and institutions.[27] This emphasis on comprehensive security is itself rooted in a positive conception of peace: 'The signatories [to the Final Act] . . . recognized that true security means more than freedom from war, that it requires economic well-being, a healthy environment and respect for human rights and fundamental freedoms,' Lamberto Zannier, the OSCE Secretary-General, observed in 2015.[28] There is explicit recognition of the linked nature of these three dimensions,[29] further underscoring the comprehensive approach, even if in actual practice more weight has often been given to the first and third dimensions. The comprehensive approach is also reflected in the organization's field operations, which in many cases have a comprehensive mandate.

Increasingly, with the end of the Cold War and the concomitant preoccupation with security risks arising from the growing number of intra-state conflicts on the European continent, the OSCE has focused a considerable amount of its attention on post-conflict peace- and state-building—or 'post-conflict rehabilitation' as the organization prefers to call it—which it now regards as a 'core task' of the organization.[30] Here, too, the organization has adopted a broad and comprehensive approach, reflected in its efforts to build capacity in such diverse areas as arms control, judicial reform, electoral regulation, human rights (including national minorities—a distinctive focus of the OSCE), policing, education, economic reconstruction, and the rule of law.

---

[27] Conference on Security and Cooperation in Europe, *Helsinki Final Act*, 1975. See also Ministerial Declaration on the OSCE Corfu Process, Athens, 2 December 2009, OSCE Doc. No. MC.DOC/1/09.

[28] Lamberto Zannier, 'Reviving the Helsinki Spirit: 40 Years of the Helsinki Final Act', *Security Community* 1 (2015), 16.

[29] OSCE Secretariat, 'The OSCE Concept of Comprehensive and Co-operative Security', OSCE Doc. SEC.GAL/100/09, 17 June 2009.

[30] OSCE Secretariat, 'Background Brief: OSCE Activities and Advantages in the Field of Post-Conflict Rehabilitation', OSCE Doc. SEC.GAL/76/11, 28 April 2011.

From Conception to Practice   41

The challenge of peacebuilding for the OSCE, as with many other peacebuilding organizations, is perceived fundamentally by the organization to be one of strengthening capacity—whether it is technical capacity, institutional capacity, or civil society capacity, etc.— often without regard, however, for the more *political* dimensions of peace and peacebuilding, notably the roles and interests of local elites, the continued existence of wartime economic and political networks, and how these elites and networks may impede efforts to build a positive peace.[31] As Jasmine-Kim Westendorf points out, 'Civil wars are, at heart, political processes. Peace processes fail when they do not respond to this central characteristic.'[32] OSCE officials are not unmindful of this important dimension but the organization is constrained by the requirement to agree mandates by consensus of the participating states, including the host country, which may object to the activities of a mission and thus impair its ability to operate effectively.[33] These constraints thus often reflect not the limits of ingenuity but, rather, the force of political reality.

North Atlantic Treaty Organization

As with both the UN and the OSCE, NATO subscribes to a comprehensive conception of peace with regard to peacebuilding activities. This comprehensive conception is reflected in NATO's descriptions of peacebuilding, including in its *Operations Assessment Handbook*, where peacebuilding is defined as 'efforts undertaken by the international community to help a war-torn society create a self-sustaining peace' that 'can include activities such as the promotion of a culture of justice, truth and reconciliation, capacity-building and promotion of

---

[31] Mats Berdal and Dominik Zaum, 'Power after Peace', in Mats Berdal and Dominik Zaum (eds), *Political Economy of Statebuilding: Power after Peace* (Abingdon: Routledge, 2013).

[32] Jasmine-Kim Westendorf, *Why Peace Processes Fail: Negotiating Insecurity after Civil War* (Boulder, CO: Lynne Rienner, 2015), 4.

[33] Interviews with OSCE officials, Vienna, November 2014. See also Walter Kemp, 'The OSCE and the Management of Ethnopolitical Conflict', in Stefan Wolff and Marc Weller (eds), *Institutions for the Management of Ethnopolitical Conflicts in Eastern and Central Europe* (Strasbourg: Council of Europe Publishing, 2008), 142.

good governance, supporting reform of security and justice institutions and socioeconomic development'.[34]

Despite its recognition of the wide range of actions that may be necessary to build a sustainable peace, NATO—unlike both the UN and the OSCE—concentrates its own efforts on a narrower range of actions, notably those that entail the use of military assets, where it enjoys a comparative advantage and perceives that it has a 'niche role' to play relative to other security organizations.[35] NATO, as well as many of the national defence departments of its members, has accordingly tended to direct its efforts in recent years towards 'stability operations' (e.g., in Iraq and Afghanistan), which the US Department of Defense defines as 'various military missions, tasks, and activities conducted . . . in coordination with other instruments of national power to maintain or reestablish a safe and secure environment, provide essential governmental services, emergency infrastructure reconstruction, and humanitarian relief'.[36] Stability operations are seen by NATO as a vital component of global peacebuilding, although many scholars and practitioners view these operations as distinct from peacebuilding because stability, the objective, is perceived to be a precondition for sustainable peace.[37] A number of NATO member states, however, including some of the states participating in these operations, take a broad view of stability. Britain, for instance, judges stability to have been established when there are 'political systems which are representative and legitimate, capable of managing conflict and change peacefully, and societies in which human rights and rule of law are respected, basic needs are met, security established and opportunities for social and economic development are open to all'.[38] This broad notion of stability/stabilization is arguably consistent with the notion of a comprehensive peace.

As noted in Chapter 1, stability—in terms of an equilibrium—can be achieved at all points along a peace spectrum. For this reason it has

[34] North Atlantic Treaty Organization, *NATO Operations Assessment Handbook*, Interim Version 1.0, 29 January 2011, 7.2.1.

[35] Interviews with NATO officials, Brussels, January 2015.

[36] US Department of Defense, 'Department of Defense Instruction (Subject: Stability Operations)', Number 3000.05, 16 September 2009.

[37] Philipp Rotmann and Léa Steinacker, *Stabilization: Doctrine, Organization and Practice*, Global Public Policy Institute, December 2013, 5.

[38] UK Foreign and Commonwealth Office, Ministry of Defence, and Department for International Development, *Building Stability Overseas Strategy*, London, 2011, 5.

From Conception to Practice   43

been criticized on conceptual grounds as devoid of meaningful content. It has also been criticized on normative grounds to the extent that it has been interpreted to mean the reinforcement of existing structures of power.[39] While NATO, it is true, has sometimes worked to support the powers that be—in both Iraq and Afghanistan, for instance—it is also true that it has sometimes challenged the prevailing powers when they have been perceived to be impediments to building peace. In Bosnia and Herzegovina, for instance, the NATO-led Stabilization Force challenged hard-line Serbs, facilitating the emergence of a more moderate Bosnian Serb leadership.[40] It seems fair to say, therefore, that there is nothing inherently conservative about building stability any more than there is anything inherently progressive about building peace.

African Union

The African Union (AU) is a regional security organization on whose continent the largest number of intra-state conflicts have erupted in the post-Cold War period. Established in July 2000 as the successor to the Organization of African Unity, the AU is a relative newcomer to post-conflict peacebuilding, although Africa's involvement in peacebuilding certainly predates it. With the establishment of the AU, however, and the adoption in July 2002 of the *Protocol Relating to the Establishment of the Peace and Security Council*, African states agreed to put in place a more formal peace and security architecture than had existed previously.[41]

The Peace and Security Council (PSC), a 'standing decision-making organ for the prevention, management and resolution of conflicts', was created *inter alia* to 'promote and implement peacebuilding and post-conflict reconstruction activities and to consolidate peace and prevent the resurgence of violence'.[42] Towards that end,

---

[39] Rotmann and Steinacker, *Stabilization: Doctrine, Organization and Practice*, 52, fn. 89.

[40] Carl Bildt, *Peace Journey: The Struggle for Peace in Bosnia* (London: Weidenfield and Nicholson, 1998), ch. 12.

[41] For a discussion of the African peace and security architecture, see Benedikt Franke, *Security Cooperation in Africa: A Reappraisal* (Boulder, CO: Lynne Rienner, 2009).

[42] African Union, *Protocol Relating to the Establishment of the Peace and Security Council of the African Union*, Addis Ababa, 2002, Articles 2(1) and 3(c).

44 *Measuring Peace*

the AU adopted a framework for post-conflict reconstruction and development (PCRD) in July 2006 whose purpose is to guide the work of the PSC and all governmental and non-governmental bodies engaged in post-conflict peacebuilding on the continent.[43] The framework does not articulate an explicit conception of peace but it is possible to distil such a notion from the activities that it identifies as contributing to sustainable peace.

The framework recognizes that post-conflict reconstruction and development is a 'holistic process' alongside early warning, conflict prevention, conflict resolution, and peace support operations. It acknowledges the importance of a 'broad range of activities' that represent the 'constitutive elements of a PCRD framework', notably in the areas of security; political governance and transition; human rights, justice, and reconciliation; humanitarian assistance; reconstruction and socio-economic development; and gender. Clearly these activities are suggestive of a much broader conception of peace than the mere absence of violent conflict. Further confirmation of this can be found in the framework's recognition of the need to address the 'root causes of conflict'. For states that historically have been extremely defensive of their national sovereignty and the principle of non-interference in domestic affairs, as many African states have been, this is a significant development. This is not to suggest, however, that practice mirrors doctrine in all respects.

Africa is notable for the large number of sub-regional organizations engaged in peace and security operations on the continent, most prominently the Economic Community of West African States (ECOWAS). ECOWAS too subscribes to a comprehensive notion of peace, or rather human security, as reflected in its Conflict Prevention Framework.[44] In another significant departure from past practice, rhetorically at least, ECOWAS privileges human security over regime security, specifically 'the creation of conditions to eliminate pervasive threats to people's and individual rights, livelihoods, safety and life; the protection of human and democratic rights and the promotion of human development to ensure freedom from fear

---

[43] African Union, *Draft Policy Framework for Post-Conflict Reconstruction and Development (PCRD)*, Addis Ababa, 2006.

[44] Economic Community of West African States, *The ECOWAS Conflict Prevention Framework*, Regulation MSC/REG.1/01/08 (2008).

From Conception to Practice

and freedom from want'.[45] ECOWAS's recognition of the structural or systemic factors that may underlie conflict—'poverty, exclusion, gender and political/economic inequalities'—also speaks to a broad view of the characteristics of and requirements for a sustainable peace.[46]

World Bank

The World Bank is concerned with peacebuilding but largely in an indirect manner. As a development bank its primary objective in low-income countries, whether conflict-affected or not, is to promote economic growth and reduce poverty. Thus, for instance, the Poverty Reduction Strategy process, the Bank's major aid framework for low-income countries since 1999, has been applied widely to conflict-affected countries, but not with the primary aim of consolidating peace in those countries.[47] However, to the extent that low economic growth, poverty, unemployment, and other related problems may be the legacies of violent conflict, as they often are, and constitute obstacles to achieving a consolidated peace, Bank programmes can be said to contribute to building peace by seeking to address these underlying problems.[48]

Although it is concerned largely indirectly with peacebuilding, the Bank (and, to a lesser extent, the International Monetary Fund) often establishes the economic and financial parameters for post-conflict reconstruction, which in turn have significant implications for the nature of the peace that is and can be built in a conflict-affected country. The Bank's conceptions—if not of peace then of associated goals such as good governance—are important because they inform the work of the Bank in ways that have consequences for peace-building. As Susan Woodward observes, 'While [the Bank and IMF] operate in a dense organizational environment and one of their

---

[45] *ECOWAS Conflict Prevention Framework*, Section II (6).

[46] *ECOWAS Conflict Prevention Framework*, Section III (10).

[47] Masatomo Nao Yamaguchi, 'Poverty Reduction Strategy Process in Fragile States: Do the PRSPs Contribute to Post-Conflict Recovery and Peace-Building in Sierra Leone?' *Journal of International Development and Cooperation* 14:2 (2008), 68.

[48] More recently the Bank has also been stressing the importance of the *prevention* of conflict as a contribution to development progress. See United Nations and World Bank, *Pathways for Peace: Inclusive Approaches to Preventing Violent Conflict* (Washington, DC: World Bank, 2018).

46   *Measuring Peace*

primary contributions is to facilitate others' financing of external
assistance to post-conflict countries, their defining role on the strategy for post-conflict reconstruction, the framework for donors, and
their conditionality-imposed concept of the state make their role
particularly consequential'.[49]

The Bank and the Fund typically figure among the first ranks of
external actors to engage with conflict-affected countries post-
conflict. In the case of Sierra Leone, for instance, the Bank and the
Fund helped to design an Interim-Poverty Reduction Strategy Paper
(I-PRSP) with the government of Sierra Leone within months of the
civil war ending in late 2000.[50] The I-PRSP was one of the earliest
comprehensive plans for Sierra Leone's post-war national development. It placed the emphasis on three priorities: the establishment of
a secure environment, the relaunching of the economy, and the
provision of basic social services to the most vulnerable groups.[51]
While there were many different constitutive components of Sierra
Leone's I-PRSP, the economic management elements can be said
broadly to have supported liberal economic policies of macro-
economic stabilization, enhanced revenue (tax) collection, reduction
of the external debt burden, and privatization and promotion of the
private sector. The Bank's liberal economic agenda, critics have
argued, has in some instances been at odds with the requirements
for peace, favouring low inflation over employment creation, for
instance, when the latter has been considered to be important for
generating support for peace consolidation.[52]

The Bank is more inclined to frame its analysis and policies in
terms of resilience, rather than peace, and seeks therefore to build

---

[49] Susan L. Woodward, 'The IFIs and Post-Conflict Political Economy', in Berdal
and Zaum (eds), *Political Economy of Statebuilding*, 154–5.

[50] Government of Sierra Leone, *Interim Poverty Reduction Strategy Paper*, Freetown, June 2001. PRSPs are produced in consultation with the national government
and civil society organizations but, as Woods observes, 'who participates and why in
IMF and World Bank consultations' is 'very selective'. Ngaire Woods, *The Globalizers:
The IMF, the World Bank and Their Borrowers* (Ithaca, NY: Cornell University Press,
2006), 171.

[51] Government of Sierra Leone, *Interim Poverty Reduction Strategy Paper*, ch. 3.1.

[52] Michael Pugh, 'The Political Economy of Exit', in Richard Caplan (ed.), *Exit
Strategies and State Building* (New York: Oxford University Press, 2012), 280–1. See
also Graciana del Castillo's critique of 'development as usual' in relation to post-conflict
economic reconstruction: Graciana del Castillo, *Rebuilding War-Torn States: The
Challenge of Post-Conflict Economic Reconstruction* (New York: Oxford University
Press, 2008).

From Conception to Practice 47

state institutions and to promote state practices that can enable conflict-affected states to withstand better the stresses that make them prone to renewed violence.[53] As a concept, resilience is more dynamic than peace, emphasizing the capacity for societies to adapt. But unless conceived explicitly in terms of the ability to maintain peace, resilience may not necessarily be convergent with peace. Jon Kurtz observes, for instance, with respect to development strategies in Uganda's Karamjoa region, that government policies that have sought to enhance the security of pastoralists—by promoting disarmament and the settlement of communities—may have undermined their long-term resilience: 'Forced disarmament and settlement of pastoralists have curtailed their ability to migrate with livestock—a vital coping strategy during drought.'[54] To the extent that it may be possible to incorporate 'conflict sensitivity' into notions of resilience, this would not appear to be an insurmountable difficulty.

International Alert and African Centre for the Constructive Resolution of Disputes

NGOs are involved increasingly in post-conflict peacebuilding, and some NGOs have a presence in the field that dwarfs that of donor states and multilateral organizations. The size of their physical footprint aside, many NGOs also contribute significantly to intellectual debates and to the development of public policy relevant to peacebuilding.

International Alert, based in London, is one of the world's largest peacebuilding organizations. It has been engaged in efforts to build peace in conflict-affected communities across the globe for more than twenty years. While it cannot be said to be representative of all NGOs working in this area, it is certainly one of the more influential. The African Centre for the Constructive Resolution of Disputes (ACCORD) similarly is one of Africa's leading conflict-management organizations. ACCORD is a South Africa-based civil society organization that conducts mediation, training, research, and conflict analysis with respect to conflicts on the African continent. It has been

---

[53] World Bank, *World Development Report 2011: Conflict, Security and Development* (Washington, DC: World Bank, 2011).
[54] Jon Kurtz, 'Resilience', United States Institute of Peace *Insights*, Summer 2014.

48   *Measuring Peace*

engaged in peacebuilding efforts in five sub-Saharan countries: Burundi, the Democratic Republic of Congo, Liberia, South Sudan, and Sudan.

Alert's work is informed by an explicit conception of peace. It is a positive conception of peace: 'the peaceful management of differences and conflict, rather than the absence of war'.[55] Alert's conception of peace is also a dynamic one: it recognizes that peace is not a fixed state, nor is it naturally self-sustaining. Peace is volatile, subject to change, especially in the case of societies emerging from conflict. What is important, Alert recognizes, is to establish whether the peace in question is tending towards or away from consolidation.

Alert defines a strong or stable peace in relation to five factors that it draws from the concept of human security. The five 'peace factors' are: power; income and assets; fairness, equality, and effectiveness of the law and legal process; safety; and mental and physical well-being. Power is of particular importance and has several aspects: voice and participation, inclusion, power differentials, social capital, leadership and legitimacy, and values and incentives. The factors are interrelated: the presence or absence of each has an influence on the others.

ACCORD's peacebuilding efforts are conducted under the auspices of its African Peacebuilding Coordination Programme. ACCORD works in support of a 'just and sustainable' peace in African countries emerging from violent conflict. Its work is also informed by a positive conception of peace expressed as:

> the development of local social institutions so that societies develop the self-sustainable and local resilience needed to manage their own tensions as well as external influences and shocks. This is what Galtung (1970) terms 'positive peace'—the causes of conflict have been removed and resilient social institutions have emerged, with the result that violent conflicts, as well as the threat of the same, are absent.[56]

Many NGOs work in the 'peacebuilding space' established by larger organizations. This is true of ACCORD, whose peacebuilding efforts have complemented those of the UN in Burundi, Liberia, and Sierra

---

[55] International Alert, *Programming Framework for International Alert: Design, Monitoring and Evaluation* (London: International Alert, 2010), 6; interviews with senior Alert officers, London, September 2014.

[56] ACCORD, *ACCORD Peacebuilding Handbook*, 2nd edn (Durban, ACCORD: 2015), 11.

From Conception to Practice   49

Leone. As is the case with many donor states, whose focus within conflict-affected states tends to be project-oriented, NGOs are more often than not concerned with the local effects of their efforts rather than with the broader strategic environment.

*   *   *

As can be seen from the foregoing discussion, there is considerable conceptual convergence among the principal peacebuilding organizations with regard to the notion of a consolidated peace. While there are certainly differences among these organizations, they tend to be matters of emphasis, often reflecting niche capabilities, rather than major substantive differences. This should not come as a surprise in view of the fact that these organizations commonly work together in support of peacebuilding in an integrated and broadly harmonious fashion. Fundamental conceptual differences, if they did exist, would likely be a hindrance to such cooperation.

Differences of other kinds also exist among peacebuilding organizations, most notably with respect to tactical considerations. These differences can have significant implications, evident for instance in the difficulties that have sometimes arisen in coordination between the military, on the one hand, and humanitarian and development organizations, on the other hand.[57] However, these differences do not have their basis in a clash of conceptions about peace. They tend instead to reflect different modes of engagement, different organizational cultures, or a lack of mutual familiarity and trust.[58]

Another important divide, and one that this chapter has not addressed, is that between international (external) and local (internal) conceptions of peace. There is considerable literature on differences between international and local perspectives on building peace after violent conflict, reflected in notions of 'friction', 'contestation', 'resistance', etc.[59] It is important not to assume that tension is inherent in

---

[57] See, for instance, Larissa Fast, 'Culture Clash: A Humanitarian Perspective on Civil–Military Relations', *Peace Policy* (April 2010), http://peacepolicy.nd.edu/2010/04/09/culture-clash-a-humanitarian-perspective-on-civil-military-interactions.

[58] Andrea Baumann, 'Clash of Organisational Cultures? The Challenge of Integrating Civilian and Military Efforts in Stabilisation Operations', *RUSI Journal* 153:6 (2008), 70–3.

[59] Representative works include Séverine Autesserre, *Peaceland: Conflict Resolution and the Everyday Politics of International Intervention* (Cambridge: Cambridge University Press, 2014); Gearoid Millar, Jair van der Lijn, and Willemijn Verkoren, 'Peacebuilding Plans and Local Reconfigurations: Frictions between Imported Processes

Measuring Peace

this relationship, although some analysts will disagree.[60] However, it is also important to recognize that divergent notions of the requirements for peace between these parties can be a serious impediment to successful peacebuilding, either because third parties misjudge local requirements or because they fail to secure the cooperation of local parties in support of peacebuilding strategies. Local perceptions are also important for measuring peace consolidation: effective local participation in assessing progress towards achieving a consolidated peace requires acceptance of a common understanding of what peace is. We turn now to an examination of assessment practices among peacebuilding actors and the extent to which local perceptions and other perspectives are factors of consideration.

and Indigenous Practices', *International Peacekeeping* 20:2 (2013), 137–43; and Oliver P. Richmond, 'Becoming Liberal, Unbecoming Liberalism: Liberal–Local Hybridity via the Everyday as a Response to the Paradoxes of Liberal Peacebuilding', *Journal of Intervention and Statebuilding* 3:3 (2009), 324–44.

[60] See, for instance, Oliver P. Richmond, *Failed Statebuilding: Intervention and the Dynamics of Peace Formation* (New Haven, CT: Yale University Press, 2014).

3

Assessing Progress

Monitoring . . . the progress of peace is of such obvious value in any post-conflict society that the worth of doing so is rarely, if ever, challenged. However the means used to do so is another matter.

Adrian Guelke[1]

How do peacebuilding actors assess progress towards the achievement of a consolidated peace? How do they know whether the peace that they are working to establish is a self-sustaining peace?

This chapter examines the dominant characteristics of the thinking and practice of major peacebuilding actors in relation to these two critical questions. Bearing in mind the difficulty of generalizing about either the thinking or the practice of these actors, this chapter offers some broad observations about how leading peacebuilding organizations take the measure of peace, to the extent that they do, and highlights particularly noteworthy practices—noteworthy for the effort that they represent to introduce greater rigour and precision into strategic assessment. Principles of good practice deriving from these and other experiences are discussed in Chapter 5.

The difficulty of generalizing about how peacebuilding organizations take the measure of peace is worth emphasizing. To begin with, many of the assessments that peacebuilding organizations conduct are concerned only indirectly with measuring peace consolidation. When the North Atlantic Treaty Organization (NATO), for instance, deployed to Bosnia and Herzegovina in 1995 tasked with implementing the military provisions of the Dayton Peace Accord, it was not the

---

[1] Adrian Guelke, 'Brief Reflections on Measuring Peace', *Shared Space: A Research Journal on Peace, Conflict and Community Relations in Northern Ireland* 18 (2014), 105.

quality of the peace that it sought to gauge. Rather, it designed a diagnostic tool to measure the 'return to normality' in Bosnia and Herzegovina. Using a number of everyday indicators of normality (e.g., the availability of food and goods, the stability of prices, the occupancy and use of public and private buildings, etc.), NATO was concerned, in its early phase of engagement, to assess progress towards achieving 'sufficient stability' in the country.[2] Similarly, the World Bank's Post-Conflict Performance Indicators (PCPIs), adopted in 2001, were concerned only tangentially with assessing the quality of the peace: they were designed to assess 'the quality of a country's policy and institutional framework to support a successful transition and recovery from conflict'.[3] While neither system of assessment seeks expressly to measure peace consolidation, both systems arguably provide relevant information for that purpose.

Another difficulty with generalizing about the thinking and practice of peacebuilding organizations is that their tools of assessment, even when they are relevant, do not necessarily reflect uniform or consistent practice within these organizations. For instance, benchmarking, as discussed below, has been employed with increasing frequency by the United Nations (UN) in its peace operations but it is not a universal practice within the organization. Some practices may be very limited indeed. Peacebuilding organizations may devise *sui generis* frameworks of assessment that are intended for a single operation, although they may have wider application potentially. NATO, for instance, in the five-year period from 1995 to 2000 in Bosnia and Herzegovina, developed and deployed three unique systems of strategic assessment (including the aforementioned) corresponding to different phases of its engagement.[4] For these and other related reasons, therefore, it is important to exercise caution when generalizing about the practices of peacebuilding organizations.

---

[2] Nicholas J. Lambert, 'Measuring the Success of the NATO Operation in Bosnia and Herzegovina 1995–2000', *European Journal of Operational Research* 140 (2002), 459–81.

[3] The original PCPI framework is presented in International Development Association, 'Adapting IDA's Performance-Based Allocations to Post-Conflict Countries', May 2001, mimeo. Later adaptations followed.

[4] William J. Owen and Stephan Flemming, 'Perspectives on the NATO Success Measurement Systems: The Record and the Way Forward', Workshop Proceedings from the Cornwallis Group VII: Analysis for Compliance and Peace Building, Ottawa, Canada, 25–8 March 2002, http://www.ismor.com/cornwallis/workshop_2002.shtml.

GENERAL OBSERVATIONS

Rare is it that peacebuilding organizations are actually tasked with assessing the robustness of the peace that they are helping to build. Reporting on developments in the field of operations is a widespread practice but reporting itself is not evaluation. And while reports may contain elements of evaluation, the evaluations are not usually concerned with measuring peace consolidation. Every head of a UN peacekeeping operation, for instance, submits formal, periodic reports to the Secretary-General on the situation in the country in question, often on a quarterly or semi-annual basis.[5] The Secretary-General, in turn, reports on the situation to the Security Council.[6] These reports vary in their content but they typically provide an overview of major developments since the last report, an evaluation of progress in the implementation of the mandate of the operation, and a discussion of the key challenges facing the country and the operation. Other peacebuilding organizations have similar reporting requirements and practices. In addition to periodic reports, heads of peace operations may also be requested to brief member-state representatives (e.g., on the UN Security Council) and donor government ministers directly, either at headquarters or in the field.

Historically, and still to this day, reporting has tended to be ad hoc and rather impressionistic. The UN Secretariat lacks a uniform understanding of the purpose of a field report and a Security Council briefing, and little guidance is provided in preparing them.[7] As a result, it is generally left to the head of the peace operation to determine how much effort it puts into information-gathering and how that is carried out, with implications for how rigorous and evidence-based the reporting will be. Some heads of operation are strongly evidence-oriented and may use public opinion surveys, small-group meetings, and benchmarking, among other empirical investigatory techniques, to inform their judgements. Heads of operation also rely on their own 'ears' and a 'gut sense' of the situation, as one former special representative of the UN Secretary-General

---

[5] There are also regular, internal reports to headquarters furnished on a monthly, weekly, or even daily basis.

[6] These reports are available at http://www.un.org/en/sc/documents/sgreports.

[7] The observations in this paragraph are drawn from author interviews with former heads of UN peace operations, 2014–16.

54   *Measuring Peace*

interviewed for this study explained, which may be buttressed by more objective evidence to varying degrees.

Other standard operating procedures in peacebuilding organizations offer the possibility for—but often fail to produce—strategic assessments of the kind that concern us here. Results-based management and results-based budgeting are widespread practices employed by the Organization for Security and Co-operation in Europe (OSCE), the UN, donor agencies, and non-governmental organizations engaged in peacebuilding that work from clearly defined objectives to articulate desired outcomes, identify relevant performance indicators, and specify resource allocation.[8] These practices are an outgrowth of the 'new public management' reform efforts of the 1990s/2000s that sought to improve public-sector performance measurement and, by extension, public-sector accountability.[9] Constructive though these reform efforts have been in many respects, they are concerned fundamentally with evaluating the performance of peace operations, not with evaluating the quality of the peace that these operations are helping to produce. Although the latter may be one measure by which the former is assessed, the converse is not true.

Notwithstanding the improvements in evaluation practice that these reforms have spurred, results-based management and budgeting and other planning and assessment practices continue to suffer from problems that have negative implications for measuring peace consolidation. For instance, as the UN's Office of Internal Oversight Services has observed with respect to UN peacekeeping operations, there is a tendency to measure what can be measured easily (e.g., the number of training sessions on human rights organized for police officers) rather than their qualitative effects (e.g., how respectful the police are of human rights)—in other words, outputs rather than outcomes.[10] More broadly, these approaches emphasize technocratic,

---

[8] For an overview of these practices, see Organisation for Economic Co-operation and Development, *Results Based Management in the Development Co-operation Agencies: A Review of Experience*, OECD/DAC Working Party on Aid Evaluation (Paris: OECD, 2000).

[9] Michael Lipson, 'Performance under Ambiguity: International Organization Performance in UN Peacekeeping', *Review of International Organizations* 5:3 (2010), 269.

[10] United Nations, Office of Internal Oversight Services, Inspection and Evaluation Division, 'Inspection Report: OIOS Review of the Relevance, Efficiency and Effectiveness of Results-Based Budgeting of Peacekeeping Operations: Improvements

Assessing Progress

quantitatively-oriented criteria to understand what are fundamentally local political processes, thus crowding out attention to local political drivers. There is also a tendency to focus narrowly on mandate implementation: i.e., how well the operation is succeeding in executing its mandate. However, mandates may not always reflect the requirements for achieving a sustainable peace. A 2010 review of UN benchmarking practices found that there were 'very few examples of benchmarks focusing on system-wide effects, and virtually no examples of benchmarks . . . focused purely on contextual aspects of progress toward sustainable peace in a country'.[11] Practice has not improved dramatically since this review was undertaken.

Meaningful assessment of conditions in a country requires robust, reliable data. However, such data is often difficult to come by in developing countries where data collection is poor, and peacebuilding operations often lack the resources needed to generate and analyse such data. The difficulty is compounded by the effects of war on a country. For years after the guns were silenced in Kosovo, the World Bank and other international actors still lacked basic economic data—including accurate population figures and data on economic activity—not only because data-collection mechanisms were not robust but also because sectarian politics impeded efforts to collect data.[12] Census data—measuring the population of respective national communities—have proved to be especially contentious throughout the Balkans because of their implications for the distribution of scarce resources. 'The more people an ethnic group can claim, the greater its pretensions to land and power,' one journalist observed in relation to census-taking in neighbouring Macedonia, where ethnic Albanians have often claimed to be under-represented.[13] Yet in the absence of basic and accurate data, it is difficult to devise

---

Needed for Realizing Full Potential', Assignment No. IED-08-002, 8 May 2008, para. 39.

[11] United Nations, *Monitoring Peace Consolidation: United Nations Practitioners' Guide to Benchmarking* (New York: United Nations, 2010), 54.

[12] For a discussion of the absence of reliable economic data in Kosovo five years after the end of armed hostilities, see Economic Strategy and Project Identification Group, 'Towards a Kosovo Development Plan: The State of the Kosovo Economy and Possible Ways Forward', ESPIG Policy Paper No. 1, Pristina, 24 August 2004, http://www.esiweb.org/index.php?lang=en&id=156&document_ID=58.

[13] 'Macedonia Census Just Inflames the Disputes', *New York Times*, 17 July 1994.

56   *Measuring Peace*

appropriate post-conflict development plans and to assess progress towards achieving objectives.

A further difficulty arises from the fact that because peacebuilding is a complex and comprehensive activity that typically engages a large number of external actors, each of whom may be engaged in specific aspects of the undertaking, these actors are often concerned primarily with performance in their particular sphere of engagement. Donor governments and coalition partners engaged in peace- and state-building in post-Taliban Afghanistan, for instance, have had responsibility for different functional areas—Italy for judicial reform, Germany for police training, the United Kingdom for counter-narcotics, and Japan for demobilization, disarmament, and reintegration[14]—and each government has been concerned especially with assessing progress in their respective functional areas, largely because they have been accountable to their legislatures for public funds deployed for that purpose. Geographic fragmentation has been a related problem: while Britain took the lead in Helmand Province, Canada was responsible for neighbouring Kandahar. Although the security, political, and peacebuilding processes were deeply intertwined, donor fragmentation meant that they were reporting on progress up separate chains—as reflected in the monitoring and evaluation reports that donor government ministries prepared for legislative review. Meanwhile, competition for funds among government agencies generates a tendency to over-report success and to be shy about reporting mixed results, failures, etc.

RECENT INNOVATIONS IN STRATEGIC ASSESSMENT

There has been growing recognition within peacebuilding organizations of the need for more rigorous and systematic analysis of the process of peace consolidation. This awareness is reflected in various innovations in strategic assessment that a number of these organizations have adopted in recent years.

---

[14] Theo Farrell, *Unwinnable: Britain's War in Afghanistan 2001–2014* (London: Bodley Head, 2017), 128.

Benchmarking (UN Peacekeeping)

Recent practice has seen the increased use of benchmarking as a mechanism within the UN system to assess progress towards the achievement of a consolidated peace. Benchmarking is a form of evaluation that uses specified standards (a value, process, or state of affairs), the attainment of which, it is thought, contributes to the realization of an operation's broad objectives. For instance, the reduction and elimination of militia threats may be a benchmark for the establishment of a secure environment.[15] The UN mission in Sierra Leone (UNAMSIL), established in October 1999 in the wake of an eight-year civil war, was the first UN peacekeeping operation to employ benchmarking to inform decision making about troop drawdown.[16] The withdrawal of the peacekeeping troops, the UN Secretary-General decided in mid-2002, should be based primarily on 'the Government's capacity to maintain external and internal security without international assistance'—a reasonable standard, one could argue, for a sustainable peace.[17] A further elaboration of benchmarks followed, notably: 1) building up the capacity of the Sierra Leone police and army; 2) completing the reintegration of former combatants; 3) consolidating the state's authority throughout the country; 4) restoring effective government control over diamond mining; and 5) making significant progress towards the resolution of the conflict in neighbouring Liberia.[18] These benchmarks were drawn from the Lomé and Abuja peace agreements and identified by their architects as elements key to the success of the peace process.

Notwithstanding the imprecision of these benchmarks, the Security Council took them on board and in July 2003 announced that it would 'monitor closely the key benchmarks for drawdown', requesting the Secretary-General to report 'on the progress made with

---

[15] For a discussion of UN benchmarking, see United Nations, *Monitoring Peace Consolidation*.

[16] UNAMSIL was the first UN peacekeeping operation but not the first UN field operation to employ benchmarking. Earlier, in 2000, the UN Mission in Bosnia and Herzegovina adopted a comprehensive Mandate Implementation Plan to guide it in completing its mandate before the closure of the mission in 2002.

[17] United Nations, 'Fourteenth Report of the Secretary-General on the United Nations Mission in Sierra Leone', UN Doc. S/2002/679, 19 June 2002, para. 42.

[18] United Nations, 'Fifteenth Report of the Secretary-General on the United Nations Mission in Sierra Leone', UN Doc. S/2002/987, 5 September 2002, paras 12–13.

58   *Measuring Peace*

respect to the benchmarks',[19] which he did on the basis of rigorous monitoring of the conditions on the ground and evaluation of potential risk factors.[20] Monitoring included tests conducted by the UN in each province of Sierra Leone to assess the readiness and capacity of government security forces to maintain peace and security independently.[21] Satisfied with the state of readiness, UNAMSIL completed its withdrawal of troops at the end of December 2005. In its reflections on UNAMSIL, the Security Council 'noted with satisfaction the innovations in UNAMSIL's methods of operation that may prove useful best practice in making other United Nations peacekeeping operations more effective and efficient, including an exit strategy based on specific benchmarks for drawdown'.[22] The Sierra Leone operation is regarded as a success to the extent at least that peace has been maintained since 2006 without reliance on external forces.

There has been further use of benchmarking in the UN system. By April 2014, seven out of seventeen UN peacekeeping operations and two out of thirteen special political missions had introduced benchmarking, often at the request of the Security Council, largely to track and measure progress.[23] Not all of the benchmarks have been concerned explicitly or exclusively with sustainable peace. They have also been used to monitor progress in relation to the rule of law, human rights, political dialogue and elections, and the extension of state authority. Moreover, benchmarks have not only been adopted for the purpose of tracking and measuring progress. They have also been used by the UN field presence to clarify an operation's objectives and thus provide direction (bearing in mind that mandates can be ambiguous); generate greater buy-in from national authorities; and promote greater accountability on the part of all parties engaged in implementing the operation's mandate.[24]

---

[19] United Nations Security Council Resolution 1492 (2003), 18 July 2003.

[20] For details of the monitoring, see United Nations, 'Eighteenth Report of the Secretary-General on the United Nations Mission in Sierra Leone', UN Doc. S/2003/663, 23 June 2003, para. 9.

[21] 'Funmi Olonisakin, *Peacekeeping in Sierra Leone: The Story of UNAMSIL* (New York: International Peace Academy, 2008), 123–4.

[22] United Nations, 'Statement by the President of the Security Council', UN Doc. S/PRST/2005/63, 20 December 2005.

[23] United Nations Department for Peacekeeping Operations and Department of Field Support, 'Concept Note: Security Council Benchmarks in the Context of UN Mission Transitions', April 2014.

[24] Ibid.

Benchmarking has the potential to bring greater rigour to measuring peace consolidation and has clearly done so in a number of cases. The benchmarking process adopted by the UN Office in Burundi (BNUB) in 2011 introduced 'surveys and measurement tools across a range of issues, including sensitive matters such as perceptions of security and other issues for which no baseline data existed', Karin Landgren, then special representative of the Secretary-General to Burundi, has observed. 'This exercise drove a productive discussion between the UN mission and the government, which ultimately signed off on the process, the indicators to be used, and the measurement methodology.'[25] Benchmarks were also used specifically to assess the progress of peace consolidation and thus inform plans for the exit of BNUB and the handover of its substantive responsibilities to the successor UN country team in 2014.[26]

Benchmarking is not without its difficulties. To be effective, benchmarks need to be concrete and meaningful. In the case of the UN's Monitoring and Tracking Mechanism of the Strategic Framework for Peacebuilding in Burundi, adopted earlier in 2007, a number of the benchmarks were vague (e.g., 'improvement of management of public resources') and unrealistic (e.g., 'existence [within one year] of a political environment conducive to the peaceful resolution of political conflict').[27] A further challenge is that the requirements for a sustainable peace may be difficult to ascertain, or they may be subject to differing interpretations, and this may have implications for analysis of progress towards achieving it. Moreover, with so much at stake—the continuation or termination of a UN peacekeeping operation, most fundamentally—it can be difficult for a peacebuilding organization to ignore entirely the political sensitivities of a host government or a leading troop or financial contributor in reporting on progress. Indeed, pressures within the UN to relegate benchmarking assessments to the annex of the Secretary-General's reports have arisen precisely for this reason.[28]

---

[25] Karin Landgren, 'Unmeasured Positive Legacies of UN Peace Operations', *International Peacekeeping* (forthcoming).
[26] United Nations Security Council Resolution 2137 (2014), 13 February 2014.
[27] United Nations Peacebuilding Commission, 'Monitoring and Tracking Mechanism of the Strategic Framework for Peacebuilding in Burundi', UN Doc. PBC/2/BDI/4, 27 November 2007, annex.
[28] Author interview with senior Department of Peacekeeping Operations official, UN Headquarters, New York, October 2014.

60   *Measuring Peace*

Conflict Analysis (OSCE)

Another instrument employed to assess the quality of the peace is conflict analysis. The OSCE has been at the forefront of peacebuilding organizations in developing a conflict-analysis 'toolkit' to assist the organization and specifically its field offices in analysing emerging threats to the peace, including post-conflict peace, while at the same time identifying entry points for conflict prevention and the building of sustainable peace. Spurred in part by the 2008 Georgia crisis, which saw an unanticipated escalation of tensions over the Georgian break-away territories of Abkhazia and South Ossetia culminate in war between Georgia and Russia, and later by ethnic violence in southern Kyrgyzstan during the 2010 revolution, the OSCE participating states took the decision in December 2011 to strengthen OSCE capabilities along all phases of the conflict cycle, including 'in early warning, early action, dialogue facilitation, mediation support and post-conflict rehabilitation on an operational level'.[29] In 2012–13, as a result, the OSCE developed internal guidelines for early warning, allowing for a more systematic and structured approach to the identification of potential conflicts and crisis situations, and rolled out a conflict analysis toolkit and a set of early warning indicators for use in the organization's field operations.[30] The Network of Early Warning Focal Points in OSCE Executive Structures was created for the purposes of information-sharing and capacity-building.

The toolkit and indicators together provide a framework for the systematic analysis of conflict settings at the local, regional, national, and international levels, including but not limited to assessments of the robustness/fragility of the peace established in the aftermath of violent conflict.[31] The framework stresses the importance of

---

[29] Ministerial Council Decision No. 3/11, OSCE Doc. MC.DEC/3/11, 7 December 2011. For a review of the process that led to the decision and the initial steps taken towards its implementation, see Alice Ackermann, 'Strengthening OSCE Responses to Crises and Conflicts: An Overview', *OSCE Yearbook 2012* (Baden-Baden: Nomos, 2013), 205–11.

[30] Author interviews with OSCE officials at the OSCE Conflict Prevention Centre, Vienna, December 2014. See also Claus Neukirch, 'Early Warning and Early Action: Current Developments in OSCE Conflict Prevention Activities', *OSCE Yearbook 2013* (Baden-Baden: Nomos, 2014), 123–33.

[31] OSCE, Conflict Prevention Centre, 'Conflict Analysis Toolkit' (draft), 4 July 2014; OSCE, 'Internal OSCE Open-Ended List of Early Warning Indicators', OSCE Doc. SEC/CPC/OS/221/12, 23 November 2012.

appreciating the relevant 'contextual issues' (e.g., socio-economic and political conditions) to comprehend the specificities of a conflict as well as the influence that the broader context can exert on conflict dynamics. The framework also emphasizes that conflict analysis is not a one-off exercise: conflicts evolve over time and analysis needs to capture these changes. It also stresses the importance of being sensitive to the drivers of peace and to giving thought to how it may be possible to sustain them. Additionally, the framework offers guidance on the use of various analytical tools (e.g., conflict profiles, conflict mapping) and visualization tools (conflict trees, conflict 'onions') that can be useful in conflict analysis. The early warning perspective focuses attention on the less visible tensions that reflect underlying fault lines and the importance of analysing these tensions 'proactively before the transformation into a violent conflict takes place'. The very extensive list of early warning indicators encompasses a wide range of potential conflict factors.[32]

The extent to which this framework has influenced actual OSCE practice has varied. By the organization's own account, 'analysis and reporting on emerging tensions and conflicts have now become more frequent, more systematic'.[33] The toolkit has been and continues to be introduced in OSCE field operations, whereas the Network of Early Warning Focal Points has helped to develop a 'corporate early warning identity' among its members. Again, by the OSCE's own reckoning, in the first year since the introduction of the framework (to July 2013), the OSCE Secretary-General raised concerns about 'worrying developments' arising from this analysis on eleven occasions with respect to the OSCE region.[34]

Officials invoke the OSCE's experience in Macedonia as evidence of the effectiveness of their efforts.[35] A combination of conflict analysis and early warning there alerted international officials to rising tensions in the run-up to the December 2016 elections, when the ruling party sought to undercut its opponents through a combination of intimidation

---

[32] The indicators are grouped beneath eight categories: 'political system, military and security structures, internal security setting, socio-economic development, environment, ethnic and religious minority groups, justice and human rights law, geopolitical situation' with numerous sub-categories and indicators associated with each one.

[33] Neukirch, 'Early Warning and Early Action', 125.   [34] Ibid., 126.

[35] Author interviews with OSCE officials in Vienna and Skopje, November 2017.

62   *Measuring Peace*

and coercion directed towards non-governmental organizations, the media, and opposition political parties. The OSCE produced trenchant analysis of the situation based on its conflict-sensitive 'Pre-Election Campaign Contextual Monitoring' that it shared with the European Commission, resulting in the latter sounding the alarm in its November 2016 progress report, where it referred sharply to a political crisis in the country manifesting itself in 'state capture', a 'divisive political culture', and a 'lack of capacity for compromise'.[36]

OSCE officials point to the quality of reporting on Ukraine as further evidence of the value of their analysis. The OSCE Special Monitoring Mission to Ukraine gathers information and reports each day on the security situation, in response to specific incidents on the ground with spot reports, and on conditions of concern with periodic thematic reports.[37] But the Ukraine experience also demonstrates one of the limitations of OSCE effectiveness. Because the OSCE is an organization whose decisions are taken by consensus, the Secretariat and especially its field operations walk a tightrope between drawing attention to emerging crises, including the eruption or re-ignition of violent conflict, and participating-state sensitivity to critical reporting. Ukraine held the OSCE chairmanship in 2013. (The OSCE chairmanship is held for one calendar year by the OSCE participating state designated as such by the OSCE Ministerial Council.) The OSCE project co-ordinator in Ukraine picked up on relevant early warning signs of the evolving crisis in autumn 2013 but had no mandate for political reporting. As one OSCE official interviewed for this study observed, 'Whenever we have dared to overstep, we have been slapped on the hand. A host can block our budget or close the mission.'[38] In addition, Ukraine made it clear that it did not want to see any OSCE involvement at that stage. Hence, the early warning mechanisms developed and put in place after 2011 could not be used to their full extent. Before the deployment of the Special Monitoring Mission, the OSCE had no effective 'eyes and ears' on the ground in Ukraine to comprehensively report on the rising tensions that culminated in the violent protests of February 2014 and the subsequent crisis over Crimea and the Donbass.

---

[36] European Commission, 'The former Yugoslav Republic of Macedonia 2016 Report', Brussels, EC Doc. SWD(2016) 362 final (9 November 2016), 1.2.

[37] Reports are available at http://www.osce.org/ukraine-smm/daily-updates.

[38] Author interview with senior OSCE official, Vienna, December 2014.

Operations Assessment (NATO)

The concept of strategic assessment is at the heart of NATO military doctrine. However, NATO's use of the term pertains to the 'comprehensive analysis of an emerging or potential crisis' leading to '[s]ound recommendations on NATO's overall aim, desired endstate, strategic objectives and desired strategic effects'.[39] Strategic, in this sense, refers to ultimate aims and objectives. The NATO term that best approximates the sense of strategic assessment employed in this study is 'operations assessment at the strategic level', which NATO defines as involving 'varying combinations of: continual measurement of strategic effects and progress towards the achievement of objectives in a military context; continual measurement of strategic progress and results in non-military domains; measurement of strategic progress and results of activities of non-military organisations; [and] an overall evaluation of progress towards the NATO end-state'.[40]

Peacebuilding organizations may employ numerous systems of strategic assessment in the course of their engagement in a conflict-affected country, and this has been particularly true of NATO. As noted at the outset of this chapter, NATO devised three distinct systems of strategic assessment in Bosnia and Herzegovina, while in Afghanistan, in the period from 2009 to 2012, NATO developed and deployed four different assessment paradigms, each one associated with a new commander of the NATO-led International Security Assistance Force (ISAF).[41]

ISAF was concerned only indirectly with building peace in Afghanistan. Its purpose initially was to assist in the establishment and maintenance of a secure environment in the capital city of Kabul and its surrounding area, gradually expanding its scope of operations to the rest of the country.[42] What is interesting for the purposes of this study, however, is one approach that ISAF introduced in 2012 for

---

[39] North Atlantic Treaty Organization, *NATO Operations Assessment Handbook*, Interim Version 1.0, 29 January 2011, 2–2, 2–4.

[40] Ibid., 2–5.

[41] Jonathan Schroden, 'Operations Assessment at ISAF: Changing Paradigms', in Andrew Williams, James Bexfield, Fabrizio Fitzgerald Farina, and Johannes de Nijs (eds), *Innovation in Operations Assessment: Recent Developments in Measuring Results in Conflict Environments* (The Hague: NATO Communications and Information Agency, 2014), 44.

[42] UN Security Council Resolution 1386 (2001) established ISAF on 20 December 2001. NATO assumed responsibility for the force in August 2003.

assessing progress towards achieving its campaign and strategic goals—an approach that arguably has utility for the purpose of measuring peace consolidation.[43] Among several noteworthy features of this approach was its reliance on inputs from ISAF subordinate/supporting commands in the field, who were thought to offer valuable perspective on local conditions. These commands were instructed to evaluate local conditions with respect to four domains (security, governance, socio-economic conditions, and regional relations), employing a set of five common standards. Security in a given area, for instance, was judged to be 'not secure'; 'partially secured but with significant risk of reversion'; 'partially secured but with moderate risk of reversion'; 'partially secured but with minimal risk of reversion'; or 'fully secured with minimal risk of reversion', and scored accordingly. Commands were instructed additionally to provide narrative justification for their choice of standards as well as a narrative overall assessment, thus combining quantitative metrics with qualitative inputs. The individual assessments were then aggregated into a composite assessment and plotted on a radar chart (see Figure 3.1). Similarities and differences in assessment among the various commands (eight in this case) were thus visible at a glance.

The foregoing practices were adopted to assess progress with respect to ISAF's campaign goals. To address the larger question of whether ISAF was achieving its overall strategic goals and objectives in Afghanistan, ISAF developed a further assessment exercise. For this purpose ISAF generated a set of core strategic questions that it distributed to ISAF staff at various levels.[44] The actual questions are classified but Schroden tell us that they would have been similar to the kinds of questions that would have been asked of the commander of ISAF at a Congressional hearing: e.g., 'Can the Afghan National Security Forces secure the country?'. The aim, Schroden explains, was to use the questions to create 'a "healthy tension" within the ISAF staff that challenged assumptions, fostered new ideas, and

---

[43] The following section draws on Schroden, 'Operations Assessment at ISAF'.

[44] Schroden observes: 'the AAG [Afghan Assessment Group] decided that the answers should be in narrative form with supporting data . . . the staff was instructed not to submit assertions in the absence of supporting facts and to ensure that answers were logically-reproducible, meaning the reader should be able to use the data or information provided to arrive at roughly similar conclusions', Ibid., 58.

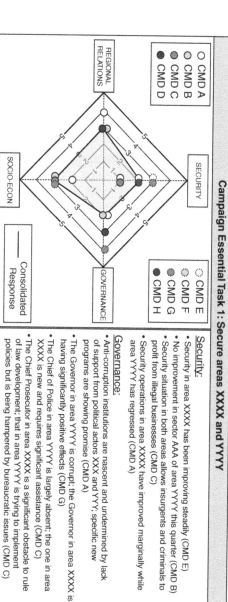

Figure 3.1. ISAF notional assessment summary slide for one campaign task

Source: Schroden 2014

identified critical issues that required attention'.[45] ISAF senior staff in turn discussed the responses to the strategic questions and used them as the basis for the production of a quarterly strategic assessment report. ISAF also organized a periodic Commander's Assessment Conference for the presentation and discussion of the findings among ISAF senior staff.

Despite its apparent strengths, the ISAF assessment approach described above did not survive for very long as the preferred approach. As one NATO operational analyst interviewed for this study explained, the narrative descriptive assessment was vulnerable to 'cherry-picking' in support of overoptimistic assessments at higher echelons of authority.[46] In his view, only an independent assessment can counter this tendency—a point that we return to in Chapter 5. What has endured, however, is recognition of the value of ground-level perspectives on progress towards achieving strategic objectives. Of course, the quality of reporting from the field can—and, in the case of Afghanistan, did—vary considerably, which can have implications for the quality of strategic assessment.

Earlier NATO assessment exercises in Afghanistan suggest additionally the value of drawing on the insights of subject matter experts whose knowledge of local languages, norms, history, and practices can help to ensure depth and soundness of judgement. The ISAF Strategic Assessment Capability (ISAC) project, launched in 2008 to feed into decision-making at NATO's strategic headquarters SHAPE, relied in part for its strategic assessment on a multi-disciplinary group of independent external experts (academics and practitioners) who contributed their analysis of conditions in Afghanistan relating to security, governance, development, and the regional context.[47] As one might expect, there may not always be a consensus among experts whose expertise, therefore, is also vulnerable to instrumental manipulation. In the case of ISAC, the problem was, rather, that the expert assessments 'routinely provided a much less positive outlook than those provided by HQs'.[48] Consequently ISAC discontinued its security domain assessments in 2011.

---

[45] Ibid.   [46] Email exchange with the author, 9 January 2015.

[47] Nick Lambert, Phil Eles, and Bruce Pennell, 'ISAF Strategic Assessment Capability: A retrospective' in ISAF Strategic Assessment Capability: Final Workshop, 10–12 December 2014 (2015), 14–25.

[48] Ibid., 20.

Assessing Progress

Co-assessment (UN Peacebuilding)

Who is best able to judge progress towards achieving a stable peace—third parties engaged in peacebuilding efforts or the conflict-affected communities? When in 2007 the UN Peacebuilding Commission adopted its integrated peacebuilding strategy for Burundi (the 'Strategic Framework'), it assigned a prominent role for monitoring and tracking progress towards achieving a consolidated peace not only to the government of Burundi but also to a wide range of Burundian civil society organizations—including representatives from opposition political parties, women's associations, the private sector, the Catholic Church and Islamic community, the journalists' association, and citizens' groups, among other non-governmental bodies.[49]

Leaving aside the merits/demerits of the UN's strategic framework and its associated monitoring and tracking mechanism,[50] the inclusion of these various stakeholders made a palpable positive difference in the process of assessing progress in at least two respects. First, it put pressure on the government—pressure that the UN might otherwise have been reluctant to exert—to place sensitive political issues on the peacebuilding agenda, notably corruption; national reconciliation; and the resumption of talks with the Palipehutu-FNL (the remaining rebel movement) and its transformation into a political party.[51] Indeed, the strategic framework highlighted political issues that the principal external actor at this stage—the World Bank—was not able to address precisely because they were political. (Article IV, Section 10 of the Bank's Articles of Agreement prohibits consideration of political matters.)[52] Second, the inclusion of a wider range of non-governmental stakeholders facilitated critical assessment of the government's conduct, which again, the UN by itself might have been

---

[49] For a list of the non-governmental bodies consulted, see United Nations Peacebuilding Commission, 'Review of Progress in the Implementation of the Strategic Framework for Peacebuilding in Burundi', UN Doc. PBC/2/BDI/10, 9 July 2008, annex I.

[50] For a critical discussion see Rob Jenkins, *Peacebuilding: From Concept to Commission* (Abingdon: Routledge, 2013), ch. 3.

[51] Author interview with Peacebuilding Support Office official, New York, October 2014.

[52] 'The Bank and its officers shall not interfere in the political affairs of any member; nor shall they be influenced in their decisions by the political character of the member or members concerned. Only economic considerations shall be relevant to their decisions.'

68   *Measuring Peace*

reluctant to do especially as it is reliant on government cooperation to achieve its objectives.[53] Thus, for instance, the release of UN Peace-building Fund monies for the development of Burundi's National Intelligence Service—a state institution whose proper functioning is critical to peace consolidation—was subject to the approval of a local human rights organization tasked with monitoring progress.[54]

Co-assessment allows third parties to draw on valuable local knowledge regarding the requirements for peace stabilization and the risks that a war-torn country or territory may face that could precipitate a relapse into violent conflict. It also provides an opportunity to build vital national capacity for ongoing monitoring of the quality of peace. By delegating responsibility for monitoring progress to national actors, moreover, the practice can also help to promote a sense of partnership with relevant international actors, which is often lacking in peacebuilding.[55] Of course national actors will have their own interests—for instance, an incumbent may have an interest in remaining in power—which can have a bearing on their assessments of progress and may even cloud their judgement. Managing divergent perceptions of the current situation and what is required to move forward is a challenge that eludes easy resolution.

PEACE INDICATORS

Another approach to measuring peace consolidation is reflected in the numerous indicators of peace, stability, resilience, and the like that are produced periodically by think tanks and research institutes.[56] These include but are not limited to the Fragile States

---

[53] Critics maintain that UN criticism remained weak nonetheless. See, for instance, Jenkins, *Peacebuilding*, 76–89.

[54] Susanna P. Campbell with Leonard Kayobera and Justine Nkurunziza, 'Independent External Evaluation: Peacebuilding Projects in Burundi', March 2010, 185, https://reliefweb.int/report/burundi/independent-external-evaluation-peacebuilding-fund-projects-burundi.

[55] Anne M. Street, Howard Mollett, and Jennifer Smith, 'Experiences of the United Nations Peacebuilding Commission in Sierra Leone and Burundi', *Journal of Peacebuilding and Development* 4:2 (2008), 44.

[56] Governments also produce their own indices. Within the United States government, the two most influential indices are USAID's Fragile States Alert List and the Political Instability Task Force's index of potential state collapse. For further details

Index (FSI), the Peace and Conflict Instability Ledger, the State Fragility Index, the States of Fragility annual report, and the Global Peace Index, all of which assess and rank states on the basis of various conflict-relevant criteria.[57] In contrast to indicators of war, such as those employed by the Correlates of War Project and the Uppsala Conflict Database Program that focus narrowly on the number of battle-related deaths in a country in a given year, indicators of peace are almost always generated by combining heterogeneous sets of data. This arguably reflects the complexity of defining peace (or cognate conditions). The way an indicator of peace is constructed is thus tied intrinsically to a particular conception of peace or, at the very least, to notions of how peace, stability, etc. should be measured. Hence, the several different indicators of peace that have been developed demonstrate that there is no consensus on how peace should be conceptualized and/or measured. This accounts in part for the sometimes significant differences in the rankings of the same countries across the various indices.

An indicator is a single number that is meant to capture a complex reality.[58] Take the annual Fragile States Index (formerly the Failed States Index), which was established in 2005 with the aim of assessing states' 'vulnerability to violent internal conflict' (or their 'susceptibility to instability', in the words of the more recent 2017 report).[59] The FSI is constructed on the basis of millions of pieces of information derived from thousands of articles, reports, and data sets that are analysed using software developed specifically for this purpose. The information is used to generate scores for twelve social, economic, and political/military indicators for each of 178 countries. Indicators include demographic pressures, refugee flows, uneven economic

---

see https://www.usip.org/sites/default/files/US-Leadership-and-the-Challenge-of-State-Fragility.pdf.

[57] Comparable indices, some now historic, include the Brookings Institution's Index of State Weakness, the World Bank's Low Income Countries under Stress, Carleton University's Failed and Fragile States Index, the BTI State Weakness Index, and the Oxford Research Group's Sustainable Security Index. For a catalogue of conflict and other indicators, see International Peace Institute, 'Catalogue of Indices 2016: Data for a Changing World', 28 September 2016, https://theglobalobservatory. org/2016/09/catalogue-indices.

[58] W. F. M. de Vries, 'Meaningful Measures: Indicators on Progress, Progress on Indicators', *International Statistical Review* 69:2 (2001), 315.

[59] The Fund for Peace, *Failed States Index 2005*, http://foreignpolicy.com/2009/10/22/the-failed-states-index-2005, *Fragile States Index 2017*, http://fundforpeace.org/fsi.

development or severe economic decline, criminalization and/or delegitimization of the state, and human rights abuse. Each indicator is scored on a scale of 0–10; the total score per country being the sum of the scores of the twelve indicators. Countries are ranked on the basis of the scores that they receive: in the most recent (2017) report reviewed for this study, Finland was deemed to be the most stable and South Sudan the most fragile of states.

While considerable amounts of data may underpin a single indicator, as is the case not only with the FSI but with other indices as well, the numerical representation of such a highly complex phenomenon as peace, stability, or fragility necessarily 'strips meaning and context from the phenomenon', as Kevin Davis, Benedict Kingsbury, and Sally Engle Merry observe.[60] It does so *necessarily* because simplification (or reductionism) is precisely the objective: it is the basis of the appeal and the utility of these indicators, whose primary intended users are policy analysts and policymakers for whom the ranking or banding of countries can facilitate decision-making with respect to aid allocation, security assistance, and other aspects of foreign relations.

Indices are of questionable value, however, for the purpose of assessing the robustness of the prevailing peace (if there is one). With a few notable exceptions, indices have limited predictive value: they provide little information about the potential future trajectory of a country towards conflict relapse. What indices offer, rather, is a snapshot of a country at a given point in time. Valuable though a snapshot can be, indices often conflate risk factors that are associated with the renewal of violent conflict with factors that are simply symptomatic of (historic) violent conflict. To give a concrete example: according to the methodology manual used to code for the FSI, 'the significant external presence of . . . a UN peacekeeping mission' would garner a score of 10 for the indicator of external intervention, with 10 reflecting a country being the most at risk of collapse and violence for a given indicator.[61] While countries that

---

[60] Kevin E. Davis, Benedict Kingsbury, and Sally Engle Merry, 'Introduction: Global Governance by Indicators', in Kevin E. Davis, Angelina Fisher, Benedict Kingsbury, and Sally Engle Merry (eds), *Governance by Indicators: Global Power through Quantification and Rankings* (Oxford: Oxford University Press, 2012), 8.

[61] The Fund for Peace (2014), *CAST: Conflict Assessment Framework Manual* (2014), 4, available at http://library.fundforpeace.org/library/cfsir1418-castmanual2014-english-03a.pdf.

Assessing Progress

attract a 'significant' UN peacekeeping presence are indeed usually among the most unstable, for predictive, explanatory, and even diagnostic purposes this way of constructing a measurement is problematic. Peacekeeping is a symptom of state weakness, not a driver of state weakness. Indeed, it may be a driver of state recovery. As I discuss in Chapter 4, there are many studies in the conflict research literature that show that peacekeeping makes a positive contribution to building peace. The upshot is that while the FSI is designed, in its own words, to make 'political risk assessment and early warning of conflict accessible to policy-makers and the public at large', it comes closer to offering a profile of fragile states rather than a health check.[62]

Often embedded in indices are implicit and perhaps even unintended methodological, theoretical, and normative assumptions. With regard to the State Fragility Index (SFI), for instance, each indicator is given equal weight in the construction of the total score for a country, suggesting that each factor is of equal importance.[63] Yet it seems improbable that some aspects measured by the indicators will not have a greater bearing on a state's peace and stability than others.[64] Similarly, the constituent components of the indicators reflect ideas about the requirements for peace, stability, etc. that may be derived from theoretical arguments, observed empirical regularities, government policy analysis and prescriptions, or even ideological convictions that are often not made explicit. In the case of the SFI, the effectiveness and legitimacy of state institutions are considered to be the critical factors that underpin stability.[65] These factors are drawn from a 'new institutionalist' perspective which, as the scholars associated with the index explain in a background paper, takes the view that 'Good institutions produce good behavior and prosperous societies; bad institutions produce bad behavior and poor

---

[62] The Fund for Peace (2016), *Fragile States Index 2016*, 13, http://fsi.fundforpeace.org.

[63] Center for Systemic Peace, *State Fragility Index and Matrix 2016*, https://www.systemicpeace.org/inscr/SFImatrix2016c.pdf.

[64] By contrast, the Peace and Conflict Instability Ledger (PCIL) uses data to determine the extent to which the indicators on average influence peace and stability in a country. This is achieved by correlating the five indicators of the PCIL with data on state instability in the 1955–2003 period. These correlations in turn are used to determine how much weight is given to each of the indicators, after which the risk scores of all countries having a population of at least 500,000 is calculated.

[65] Center for Systemic Peace, 'State Fragility Index and Matrix 2016', in Monty G. Marshall and Gabrielle Elzinga-Marshall, 'Global Report 2017: Conflict, Governance, and Fragility', http://www.systemicpeace.org/globalreport.html.

72 *Measuring Peace*

societies ... [and that] changing institutions can, in relatively short order, lead to changes in behavior—for good or for ill'.[66] These assumptions will not be visible to the casual user of indices. Moreover, as Nehal Bhuta observes in an incisive study of indices of state fragility: 'Terms such as legitimacy and effectiveness are controversial, and subject to a wide range of plausible definitions deriving from different normative and sociological frameworks ... The serious difficulties of determining the effects of political institutions on political outcomes across very diverse social and historical realities are well-known.'[67] The point is not to suggest that the perspective adopted here or elsewhere is necessarily a flawed one but, rather, to draw attention to the presence of underlying assumptions and to underscore the importance of reflecting on them critically, which, again, the casual user of indices is unlikely to do. The 'problem in this approach—namely selecting indicators based on particular ... theories, narratives, perspectives or pre-assumptions—is that its use of measurements leads to consolidating (rather than challenging) the pre-assumptions', Svein Erik Stave observes.[68]

Some indices would appear to be more promising than others from the standpoint of assessing peace durability. The Institute for Economics and Peace (IEP) produces the Global Peace Index and, within it, the Positive Peace Index (PPI), which, the IEP claims, 'can be used as the basis for empirically measuring a country's resilience, or its ability to absorb and recover from shocks'.[69] The PPI is constructed on the basis of a scoring of eight domains or 'pillars' of positive peace for 162 countries: well-functioning government, sound business environment, low levels of corruption, high levels of human capital, free flow of information, good relations with neighbours, equitable

---

[66] Jack Goldstone, Jonathan Haughton, Karol Soltan, and Clifford Zunes, 'Strategy Framework for the Assessment and Treatment of Fragile States', USAID PPC/IDEAS and Center for Institutional Reform and the Informal Sector, Washington, DC, November 2003, 5.

[67] Nehal Bhuta, 'Governmentalizing Sovereignty: Indexes of State Fragility and the Calculability of Political Order' in Davis, Fisher, Kingsbury, and Merry (eds), *Governance by Indicators*, 137.

[68] Svein Erik Stave, 'Measuring Peacebuilding: Challenges, Tools, Actions', NOREF Policy Brief No. 2 (Oslo: Norwegian Peacebuilding Resource Centre, May 2011), 3.

[69] Institute for Economics and Peace, *Positive Peace Report 2016* (Sydney: IEP, 2016), 8. Recalling the discussion in Chapter 1, note the conceptualization of positive peace (i.e., 'resilience') employed here.

Assessing Progress   73

distribution of resources, and acceptance of the rights of others.[70] The eight pillars, and the twenty-four indicators used to measure them, have been selected based on the IEP's statistical analysis of 'thousands of cross-country measures of economic and social progress to determine what factors have a statistically significant association with Negative Peace'.[71] The pillars and indicators, the IEP maintains, reveal the presence or absence of the attitudes, institutions, and structures that create and sustain peaceful societies.[72]

The IEP claims that the index can be used 'to measure fragility and to help predict the likelihood of conflict, violence and instability'.[73] However, as the IEP clarifies further, 'the aim is not to predict when a shock will happen or how a country will fare after a shock, but how well equipped it is to rebound and adapt to the shocks it faces'.[74] It is not clear how precisely, then, the index can assess the durability of peace. Take Liberia, a war-ravaged country whose UN peacekeeping operation (UNMIL), established in 2003, withdrew from the country on 30 March 2018. What does the PPI tell us about the prospects of survival of peace in post-UNMIL Liberia? It is ranked 129 out of 162 countries, just below Mozambique and just above Djibouti. It has an overall score of 3.499; the top score, held by Denmark and Finland, being 1.361. One can examine the scores reported for each of the eight pillars and probe them for clues about the future trajectory but, as the authors of the report acknowledge, positive peace is a complex system and the relationship between the pillars is not a linear one.[75] Moreover, the IEP recognizes that each country's peace is unique: 'IEP does not attempt to determine the specific *attitudes, institutions and structures* necessary for Positive Peace, as these will very much be dependent on cultural norms and specific situations. What is appropriate in one country may not be appropriate in another.'[76]

Two other sets of indicators are noteworthy for their emphasis on local perspectives in measuring peace consolidation and for their heterogeneous nature. The first are the indicators associated with

---

[70] Ibid. For details of the methodology, see appendix A.   [71] Ibid.
[72] Ibid.   [73] Ibid.   [74] Ibid., 23.
[75] 'In systems thinking, the system is more than the sum of its parts and therefore cannot be understood merely by breaking it down. This contradicts the notion of linear causality in understanding the way a country operates...This is why it is important to look at the multidimensional concept of Positive Peace as a holistic, systemic framework' (Ibid., 3).
[76] Ibid., 11.

the Peace and Security Goals that have been developed in the context of the International Dialogue on Peacebuilding and Statebuilding. The International Dialogue is a forum for political dialogue among fragile and conflict-affected states, development agencies, and civil society.[77] It was born of the Organisation for Economic Co-operation and Development-facilitated New Deal for Engagement in Fragile States, which was signed in Busan, South Korea, in November 2011. As part of the New Deal, it was agreed that the g7+—a group of now twenty countries that are or have been affected by violent conflict—and international partners would develop a set of indicators to track progress in key areas of peacebuilding within each of the countries over time.[78] The key areas—the Peace and Security Goals—were identified as legitimate politics, security, justice, economic foundations, and revenue and services,[79] and a total of thirty-four global indicators have been adopted from which national indicators are selected to assess progress towards meeting these goals.[80] The International Dialogue indicators are distinctive in so far as they (and the goals they underpin) have been formulated by the conflict-affected countries themselves, reflecting the view that peacebuilding historically has not taken sufficient account of the local specificities of these countries. Those specificities are derived in part from 'fragility assessments'—inclusive and participatory exercises carried out, again by the countries themselves, to assess the causes, features, and drivers of their fragility. The g7+ countries have also taken the lead in monitoring to ensure their ownership of the evaluation process.[81]

In addition to the bottom-up process in which the International Dialogue indicators have been formulated, another crucial feature is

[77] See International Dialogue on Peacebuilding and Statebuilding, https://www.pbsbdialogue.org/en.

[78] International Dialogue on Peacebuilding and Statebuilding, 'Progress Report on Fragility Assessments and Indicators', 2012, http://www.newdeal4peace.org/wordpress/wp-content/uploads/progress-report-on-fa-and-indicators-en.pdf.

[79] 'A New Deal for Engagement in Fragile States', adopted on 30 November 2011 at the 4th High Level Forum on Aid Effectiveness, Busan, http://www.g7plus.org/en/new-deal/document.

[80] Indicators specified in International Dialogue on Peacebuilding and Statebuilding, 'Peacebuilding and Statebuilding Indicators: Progress, Interim List and Next Steps', 2013, 2, http://www.pbsbdialogue.org/documentupload/03%20PSG%20Indicators%20EN.pdf.

[81] Erin McCandless, 'Wicked Problems in Peacebuilding and Statebuilding: Making Progress in Measuring Progress through the New Deal', *Global Governance* 19 (2013), 227–48.

Assessing Progress 75

that they are not intended to rank countries. Instead they offer a baseline for reporting change within a country over time. Moreover, the indicators are neither fixed nor uniform: the list is meant to evolve over time to reflect evolving and distinct conditions within the countries. The indicators thus give some freedom to policymakers in war-affected countries in terms of how they assess progress in peacebuilding. However, such scope comes at the expense of comparability while the large number of indicators comes at the expense of parsimony, which makes it difficult to unpack the indicators into broader planning frameworks and to leverage the indicators into decisions about relative donor resource allocation—a consideration that should be a natural outcome of all these indices but rarely is.[82] Indeed, turning the International Dialogue indicators into actual policy was highlighted as a major challenge at the 4th Global Meeting of the International Dialogue held in Freetown, Sierra Leone, in June 2014, where the first New Deal Monitoring Report was presented. Another major challenge reported was the collection of data, which has proved difficult as a result of the lack of capacity on the part of many countries to collect the relevant data.[83] For these and other reasons—notably the Ebola crisis in Liberia and Sierra Leone from 2014—the indicators initiative has lost momentum.

Another promising set of indicators that emphasizes local perspectives and heterogeneity are those being generated by the Everyday Peace Indicators project led by Pamina Firchow, Roger Mac Ginty, and Naomi Levy.[84] Working with local non-governmental organizations, the project is a bottom-up approach to measuring peace that utilizes focus groups in conflict-affected communities to identify indicators that community members themselves employ to assess the quality of the peace, threats to the peace, etc. based on their life experiences. Examples of indicators that have been generated locally include: a decline in barking dogs (indicating decreasing levels of movement by soldiers or rebel fighters on the perimeter of a community), a decline in sectarian graffiti (indicating a decline in tension between opposing groups), and storeowners painting their storefronts

---

[82] *Progress Report on Fragility Assessments and Indicators*, 11.

[83] International Dialogue on Peacebuilding and Statebuilding, 'New Deal Monitoring Report 2014: Final Version', 13–14.

[84] Details of the project can be found at https://everydaypeaceindicators.org.

(indicating confidence that violence and the destruction of property will not resume).[85]

As well as reflecting local perceptions of what peace means, the Everyday Peace Indicators also reflect local conditions. Much of the data used for the purposes of conventional conflict analysis, quantitative analysis especially, is national-level data. However, war and peace are often very local phenomena, exhibiting considerable variation across a national territory. Everyday Peace Indicators offer the advantage of being able to capture variation across a territory. This strength is also a weakness as the idiosyncratic nature of the indicators militates against their use for the purpose of systematic comparison. However, the project leaders contend that the Everyday Peace Indicators are not meant to supplant other measures but, rather, to complement and add value to them.[86]

*　*　*

This chapter has examined dominant characteristics of the thinking and practice of leading peacebuilding actors with regard to assessing progress towards the achievement of a sustainable post-conflict peace. It has highlighted innovative approaches to strategic assessment that a number of these actors have developed and employed in recent years—approaches that promote more rigorous analysis of the quality of peace—its robustness, vulnerabilities, durability. It has also examined various indices and indicators of peace which are widely consulted in analysis and, to some degree, for purposes of policy planning. From these experiences it is possible to extract principles of good practice, which I will do in Chapter 5. First, however, in Chapter 4, I present the findings of research conducted with a colleague into factors of post-conflict peace stabilization. Based on a combination of quantitative analysis and case studies, the research contributes to a long-running debate about the durability of peace that is relevant to this study. The research also raises questions about how best to analyse peace durability, which I discuss in Chapter 5.

---

[85] Roger Mac Ginty, 'Indicators +: A Proposal for Everyday Peace Indicators', *Evaluation and Program Planning* 36 (2013), 56–63.

[86] Pamina Firchow and Roger Mac Ginty, 'Measuring Peace: Comparability, Commensurability and Complementarity Using Bottom-Up Indicators', *International Studies Review* 19:1 (2017), 6–27.

4

Factors of Post-Conflict Peace Stabilization

With Anke Hoeffler

All failed peaces are alike; every successful peace succeeds in its own way.

Mike McGovern[1]

This chapter, co-written with my colleague Anke Hoeffler, seeks to identify factors that contribute to the durability of post-conflict peace. It represents a further effort, alongside those identified in the Introduction, to identify the causes of enduring peace. We are interested in explaining why peace endures in countries that have experienced civil war. We evaluate the salience of a number of factors in relation to the duration of peace in all countries that have experienced peace after civil war since 1990. We use a mixed-methods approach that combines survival analysis—a statistical method—of 205 post-conflict peace 'episodes' since 1990, with analysis of six cases of post-conflict peace (Burundi, East Timor, El Salvador, Liberia, Nepal, and Sierra Leone), specially prepared for this study by country experts.[2]

This chapter may appear at first glance to be out of keeping with the spirit of this book. After all, the principal argument I make in the book is that rigorous assessments of the robustness of peace call for an appreciation of the characteristics of, and the requirements for, a

---

[1] Mike McGovern's case study for the 'Factors of Post-Conflict Peace Stabilization' project on which this chapter is based.
[2] Jeremy Allouche (Sierra Leone); Charles T. Call (El Salvador); Paul Jackson (Nepal); Mike McGovern (Liberia); Janvier Nkurunziza (Burundi); and Kate Roll (East Timor/Timor-Leste). Cases were selected in consultation with the UK Department for International Development, one of the sponsors of the research reflected in this chapter.

78 *Measuring Peace*

stable peace in a given conflict situation, and correspondingly strong knowledge about the conflict dynamics specific to that conflict situation. Such an understanding cannot be achieved on the basis of a statistical analysis of scores of peace episodes and a half dozen case studies.[3] However, statistical analysis can be useful in helping to identify potentially apposite factors of peace stabilization, while comparative case study analysis can provide insights into the causal dynamics associated—or not—with these factors.

We present our analysis below. Any reader not interested in the details of the statistical analysis can skip to the Discussion on p. 99 of this chapter for the main findings and then continue to Chapter 5.

EXISTING APPROACHES TO ANALYSIS OF PEACE DURATION

As discussed in the Introduction to this book, there is a growing body of literature that applies quantitative methods to the study of the duration of peace in the wake of civil war. The sample is typically limited to countries that have experienced at least one spell of armed conflict. This is in contrast to the onset literature, which includes countries that have never experienced armed conflict.[4] In the analysis of post-conflict countries a number of different quantitative methods can be applied. One option is to investigate whether a new war broke out and ended the peace. The endurance or breakdown of the peace can be coded as a 0/1 variable, and limited dependent variable analysis can be applied to estimate which factors affect the probability of the recurrence of war.[5] However, if one is not just interested in the

---

[3] For a discussion of the strengths and limitations of quantitative methods for studying civil war recurrence, see Charles T. Call, *Why Peace Fails: The Causes and Prevention of Civil War Recurrence* (Washington, DC: Georgetown University Press, 2012), 59–65.

[4] See, for instance, Håvard Hegre, Tanja Ellingsen, Scott Gates, and Nils Petter Gleditsch, 'Toward a Democratic Civil Peace? Democracy, Political Change, and Civil War 1816–1992', *American Political Science Review* 95:1 (2001), 16–33; James D. Fearon and David D. Laitin, 'Ethnicity, Insurgency, and Civil War', *American Political Science Review* 97:1 (2003), 75–90; and Paul Collier and Anke Hoeffler, 'Greed and Grievance in Civil War', *Oxford Economic Papers* 56:4 (2004), 563–95.

[5] See, for example, Monica Duffy Toft, *Securing the Peace: The Durable Settlement of Civil Wars* (Princeton, NJ: Princeton University Press, 2009); Joakim Kreutz, 'How

Factors of Post-Conflict Peace Stabilization 79

question of *whether* the peace breaks down but also in *how long* a peace spell lasts, then the use of survival (or duration) analysis is the appropriate choice of method. Survival analysis is a statistical method that allows researchers to analyse how long a specific state lasts until the occurrence of a specific event. It is commonly applied in medical studies where the effect of a particular treatment on the survival time of patients is evaluated. In our study we apply survival analysis to examine the impact of a number of variables on the longevity of the peace.

Although a number of studies apply duration analysis to the study of peace, there is no consensus among scholars regarding the drivers of enduring peace. Hartzell, Hoddie, and Rothchild find that the most durable settlements are those in which the civil conflict was of long duration; the previous governing regime was democratic; the peace agreement contains provisions for the territorial autonomy of threatened groups; and there are third-party security guarantees.[6] In a subsequent study, Hartzell and Hoddie extend the analysis to examine the effects of power-sharing arrangements on the duration of peace settlements.[7] They find that settlements that promise power sharing increase the likelihood that the settlement will endure. Martin extends this analysis further. He challenges the prevailing view that elite power-sharing pacts are critical for peace survival and argues that institutional options such as territorial power sharing and proportionality in military forces yield a more durable peace.[8] Nilsson, on the other hand, finds that all-inclusive peace deals—signed by the government and all rebel groups—do not necessarily yield lasting peace.[9]

Fortna's seminal work on the impact of United Nations peacekeeping operations (UNPKOs) suggests that the presence of UNPKOs

and When Armed Conflicts End: Introducing the UCDP Conflict Termination Dataset', *Journal of Peace Research* 47:2 (2010), 243–50; and Call, *Why Peace Fails*.

[6] Caroline Hartzell, Matthew Hoddie, and Donald Rothchild, 'Stabilizing the Peace after Civil War: An Investigation of Some Key Variables', *International Organization* 55:1 (2001), 183–208.

[7] Caroline Hartzell and Matthew Hoddie, 'Institutionalizing Peace: Power Sharing and Post-Civil War Conflict Management', *American Journal of Political Science* 47:2 (2003), 318–32.

[8] Philip Martin, 'Coming Together: Power-Sharing and the Durability of Negotiated Peace Settlements', *Civil Wars* 15:3 (2013), 332–58.

[9] Desirée Nilsson, 'Partial Peace: Rebel Groups Inside and Outside of Civil War Settlements', *Journal of Peace Research* 45:4 (2008), 479–95.

80 *Measuring Peace*

significantly improves the chances of peace surviving.[10] In the post-Cold War period (to 1999), she observes, UNPKOs have reduced the risk of peace breaking down by about 50 per cent. She finds that most other variables, such as the outcome of the conflict, the nature of the conflict (identity), the death toll in the conflict (intensity), the nature of the previous governing regime (degree of democracy), and the relative size of the government army are insignificant. Only the presence of UNPKOs, the duration of the conflict, and economic development are significant for maintaining the peace. Further evidence of the importance of UNPKOs in reducing the risk of renewed war is found by Hultman et al.; Mason et al.; Gilligan and Sergenti; and Collier et al.[11] Rudloff and Findley and more recent work by Walter, on the other hand, find little evidence that peacekeeping increases the length of the peace.[12] Walter concludes, furthermore, that peace spells that end with a peace agreement following territorial conflicts and include good government accountability measures (i.e., participation, written constitution, free press, rule of law) increase the likelihood of peace survival.[13] 'The more accountable the government is to a wide range of people, the easier it will be to credibly commit to share power and reform, and the fewer incentives groups will have to return to violence,' she observes.[14] None of the other variables in her analysis, including UNPKOs, income, polity measures, and the duration and intensity of the previous conflict, are significant.

[10] Virginia Page Fortna, 'Does Peacekeeping Keep Peace? International Intervention and the Duration of Peace after Civil War', *International Studies Quarterly* 48:2 (2004), 269–92; Virginia Page Fortna, *Does Peacekeeping Work? Shaping Belligerents' Choices after Civil War* (Princeton, NJ: Princeton University Press, 2008).

[11] Lisa Hultman, Jacob D. Kathman, and Megan Shannon, 'United Nations Peacekeeping Dynamics and the Duration of Post-Civil Conflict Peace', *Conflict Management and Peace Science* 33:3 (2016), 231–49; David T. Mason, Mehmet Gurses, Patrick T Brandt, and Jason Michael Quinn, 'When Civil Wars Recur: Conditions for Durable Peace after Civil Wars', *International Studies Perspectives* 12:2 (2011), 171–89; Michael J. Gilligan and Ernest J. Sergenti, 'Do UN Interventions Cause Peace? Using Matching to Improve Causal Inference', *Quarterly Journal of Political Science* 3:2 (2008), 89–122; and Paul Collier, Anke Hoeffler, and Måns Söderbom, 'Post-Conflict Risks', *Journal of Peace Research* 45:4 (2008), 461–78.

[12] Peter Rudloff and Michael G. Findley, 'The Downstream Effects of Combatant Fragmentation on Civil War Recurrence', *Journal of Peace Research* 53:1 (2016), 19–32; Barbara Walter, 'Why Bad Governance Leads to Repeat Civil War', *Journal of Conflict Resolution* 59:7 (2015), 1242–72.

[13] Walter, 'Why Bad Governance Leads to Repeat Civil War'.   [14] Ibid., 1245.

Factors of Post-Conflict Peace Stabilization   81

The qualitative and mixed-method literature is similarly inconclusive, in part because the notion of peace itself is defined variably, with some scholars working with a minimal conception of peace (absence of violent conflict) and others with more ambitious conceptions of peace (e.g., elimination of root causes of conflict or 'participatory peacebuilding'). Scholarship in this area has stressed the importance of the nature of civil war termination (Licklider), third-party security guarantees (Fortna), transparency between combatants (Doyle, Johnstone, and Orr), 'institutionalization before liberalization' (Paris), security-sector reform (Toft), and inclusive political settlements (Call), among other factors. As with the quantitative analysis, there is a lack of consensus among scholars regarding the factors underpinning peace duration.[15]

Not many variables, then, appear to be significant in the duration of peace analysis and, yet, scholars disagree about the importance of a number of them. This suggests that it is hard to explain the duration of peace in general. Indeed, as one of the case study authors for this project aptly observed:

> It has been well documented that countries that have experienced civil wars have a high probability of falling back into war . . . Yet we know less about how long a peace must last until it is likely to 'stick', and still less about how and why that dynamic pertains. For the moment, the state of our knowledge appears something like the opening of *Anna Karenina* turned on its head: 'All failed peaces are alike; every successful peace succeeds in its own way.'[16]

KEY TERMS

For our statistical analysis we first need to define precisely the key terms that are germane to the parameters of our investigation. Our definition of *post-conflict* is the absence of armed conflict, i.e., a

---

[15] See, respectively, Roy Licklider (ed.), *Stopping the Killing: How Civil Wars End* (New York: New York University Press, 1993), ch. 13; Fortna, 'Does Peacekeeping Keep Peace?'; Michael W. Doyle, Ian Johnstone, and Robert C. Orr, *Keeping the Peace: Multidimensional UN Operations in Cambodia and El Salvador* (Cambridge: Cambridge University Press, 1997); Roland Paris, *At War's End: Building Peace after Civil Conflict* (Cambridge: Cambridge University Press, 2004); Monica Duffy Toft, 'Ending Civil Wars: A Case for Rebel Victory?' *International Security* 34:4 (2010), 7–36; and Call, *Why Peace Fails*.

[16] Mike McGovern's case study for this project.

82   *Measuring Peace*

'negative' peace. Most quantitative studies of armed conflict employ a negative conception of peace, with armed conflict being defined variably depending on which data set is adopted. Many post-conflict situations in fact are not entirely peaceful but, rather, are character-ized by ongoing, sporadic violence.[17] However, if the level of violence is below the given threshold of armed conflict, we define these situations as post-conflict.

Our definition of *armed conflict* is based on the Armed Conflict Dataset (ACD) produced by the Uppsala Conflict Data Program and the Peace Research Institute Oslo.[18] It is the most widely used data set for the study of armed conflict. The most recent version of the ACD available to us starts at the conclusion of World War II and ends on 31 December 2013. Only very few armed conflicts are international conflicts—i.e., between states—and we disregard these conflicts. We focus on conflicts that are internal to a country: these conflicts may or may not receive support from beyond the national borders. The ACD coders also distinguish between 'major' and 'minor' armed conflicts. *Major armed conflicts* cause at least 1,000 battle-related deaths a year. Military as well as civilian deaths are counted as 'battle related'. A further part of the definition is that there is organized effective violent opposition to the government. This distinguishes this type of violence from genocides, pogroms, communal violence, and the like. *Minor armed conflict* is limited to 25 to 999 battle deaths per year. We define both major and minor armed conflicts as armed conflicts.

The ACD provides information by armed conflict. One example would be the FARC rebellion against the government of Colombia, where the conflict has lasted a long time and has only one conflict episode (1964–2013, i.e., ongoing at the end of the coding period) because the associated battle deaths have exceeded the armed conflict threshold each and every year in that period. The Palipehutu rebellion against the government of Burundi is listed as one conflict with four distinct episodes (1965, 1991–2, 1994–2006, 2008) because there have

---

[17] Astri Suhrke and Mats Berdal (eds), *The Peace In Between: Post-War Violence and Peacebuilding* (Abingdon: Routledge, 2012); Michael J. Boyle, *Violence after War: Explaining Instability in Post-Conflict States* (Baltimore, MD: Johns Hopkins University Press, 2014).

[18] Lotta Themnér and Peter Wallensteen, 'Armed Conflicts: 1946–2011', *Journal of Peace Research* 49:4 (2012), 565–75; Nils Petter Gleditsch, Peter Wallensteen, Mikael Eriksson, Margareta Sollenberg, and Håvard Strand, 'Armed Conflict, 1946–2001: A New Dataset', *Journal of Peace Research* 39:5 (2002), 615–37.

Factors of Post-Conflict Peace Stabilization 83

been either few or no battle deaths in the intervening periods. Other countries have experienced a number of distinct armed conflicts with one or more episodes each, e.g., Nigeria (Biafra 1967–70; Niger Delta 2004; Boko Haram 2009, 2011–ongoing). Other countries, such as Burma (Myanmar), have experienced a number of distinct conflicts at the same time (rebellions by the Karen, Karenni, Shan, Kokang, Kachin). As a unit of observation, we focus on the conflict episode, and the post-conflict episode starts when the conflict episode ends. This is irrespective of whether there is another ongoing conflict in the same country or whether this same conflict resumes at some later point in time.

Some analysts will disagree with the judgement made by the authors of the ACD data set. The 2006 violence in East Timor, for instance—one of our case studies—left thirty-eight dead and forced 150,000 to flee their homes but it is not noted by ACD, perhaps because it fails to satisfy the requirement that the opposition must be a 'formally organised opposition group'. However, the crisis is widely regarded as evidence of the failure of the peace to hold.[19] Similarly, the 1972 purges in Burundi, another of our case studies, are not captured by the armed conflict definition in the ACD data set but are considered by many analysts to be an important part of the cycle of violence.[20] Again, this is why detailed knowledge of specific armed conflicts, which case study analysis permits, is a useful complement to the statistical analysis.

In our definition, the end of the armed conflict is the beginning of the post-conflict period or peace spell. While some armed conflicts end in settlements or military victories, many conflicts continue at a lower level. ACD does not record an ongoing armed conflict if there are fewer than twenty-five battle-related deaths per year. The termination of an armed conflict is categorized by Kreutz.[21] He distinguishes between military victory, peace agreements, ceasefires, and 'other outcomes'. Victory is when one side is either defeated or eliminated, capitulates, or surrenders. A peace agreement is defined as an agreement between the main actors concerned with the resolution of the conflict and may be accepted while armed activity is ongoing.

---

[19] Kate Roll's case study for this project.
[20] Janvier Nukurunziza's case study for this project.
[21] Kreutz, 'How and When Armed Conflicts End'; Joakim Kreutz, UCDP Conflict Termination Dataset Codebook, v.2-2015, 19 February 2016.

84   *Measuring Peace*

Table 4.1. Armed conflict outcomes, 1990–2013

Outcome	Count	%
1 Peace agreement	31	15
2 Ceasefire	41	20
3 Government victory	30	15
4 Rebel victory	9	4
5 No or low activity	88	43
6 Actor ceases to exist	6	3
Total	205	100

Source: UCDP Termination Dataset version 2.0-2015 and Kreutz
2010. There are 210 conflict episodes that ended during 1990–2013
but for five observations the termination is not coded

Conflicts are coded as having terminated by peace agreement if this
agreement is followed by military inactivity. By contrast, ceasefires
are agreements that terminate military operations but do not entail a
resolution of the conflict. However, a large number of armed conflicts
do not end in either victory or settlement but 'rumble on' without
producing the required twenty-five battle-related deaths. This category makes up 43 per cent of all observations and is termed 'low or
no activity'. The remaining category is cases in which other criteria
are not met, for instance one side in a conflict ceases to exist or is
defeated in another simultaneous conflict. For the 205 conflict episodes that ended after 1989, Table 4.1 presents the frequencies for the
various outcomes.

A FIRST LOOK AT THE SURVIVAL OF PEACE

Using the ACD we focus on the post-Cold War period. Thus, we only
consider armed conflicts that ended in or after 1990; the last year we
observe is 2013. This provides us with 210 peace spells as discussed
above. Of these peace spells 62 were single-spell episodes, i.e., the
peace started and then either lasted until the end of the period or
ended due to conflict that lasted until 2013. The other 148 peace spells
are multiple spells in which the conflict recurred, then ended, and at
least one further spell of peace was observed.

Before turning to the regression analysis, we want to examine the
empirical patterns of the peace spell data: how many peace spells

Factors of Post-Conflict Peace Stabilization   85

break down and when does this happen? This information is provided by the Kaplan-Meier survival estimates as shown in Figure 4.1 and Table 4.2. Figure 4.1 shows peace spells measured in days. In the beginning, all of our observations are at peace and as time passes, some peace spells come to an end and some continue. Following from the ACD data definition, conflicts are defined by a minimum of twenty-five battle-related deaths per year and a peace period cannot

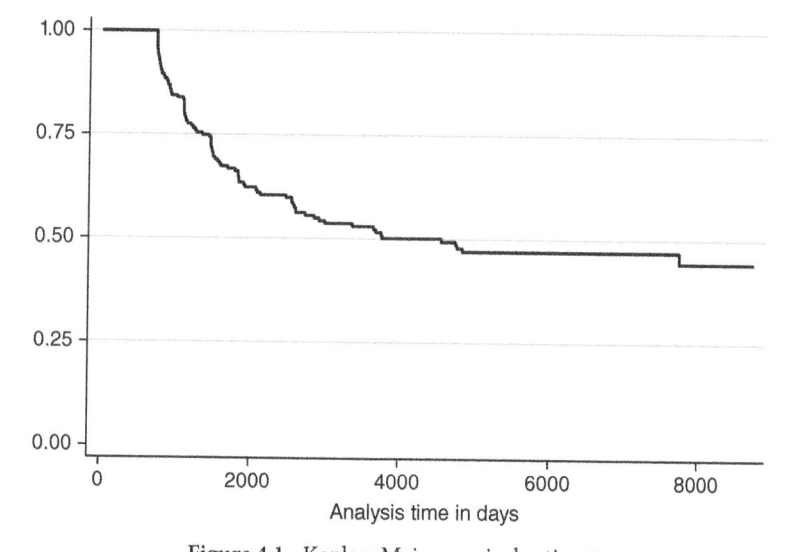

Figure 4.1. Kaplan-Meier survival estimate

Table 4.2. Number of peace spells surviving

End of year	No. of peace spells	Fail	Survivor function (%)
1	205	0	100
2	201	1	99.5
3	160	33	82.8
4	141	16	74.4
5	119	15	66.3
6	102	10	60.5
7	99	1	59.9
8	88	8	55.0
9	83	2	53.7
10	77	1	53.0
11	71	4	50.2
12	68	0	50.2

86 *Measuring Peace*

be shorter than one year; this accounts for the first flat part of the Kaplan-Meier graph. From the end of the first year until approximately 5.5 years (2,000 peace days) the survivor estimates drop more sharply than after. This suggests that peace spells are more likely to break down within the first five years than in the following five years. Table 4.2 provides the same information. After two years 99.5 per cent of all peace spells survive, i.e., half of 1 per cent of the peace spells have failed (war recurred). After three years 82.8 per cent of the peace spells have survived. After twelve years only about half of the peace spells have survived (50.2 per cent).

Figure 4.2 graphs the survivor functions by outcome of the armed conflict. We distinguish among three different outcomes: settlement (peace agreements and ceasefires combined), victory (government or rebel victory), and other (low activity or actor ceases to exist). Higher lines represent longer survival, i.e., a lower risk or hazard of failure (armed conflict breaking out again). According to Figure 4.2, victories are associated with longer peace spells, followed by settlements, while peace spells after low activity are most likely to break down.

In Figure 4.3 we graph the peace spells with UNPKOs and without. UNPKOs include some special political missions and are led by the UN Department of Peacekeeping Operations. We define UNPKO as a

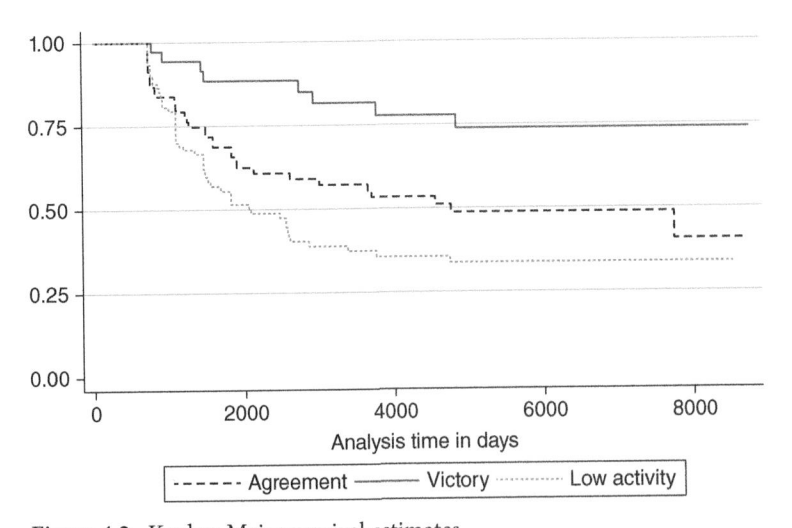

Figure 4.2. Kaplan-Meier survival estimates

Note: outcome = 0 refers to settlement, outcome = 1 refers to victory, and outcome = 2 refers to 'other'. Log-rank test for equality of survivor functions chi2(2) = 15.96 Pr > chi2 = 0.0003

Factors of Post-Conflict Peace Stabilization

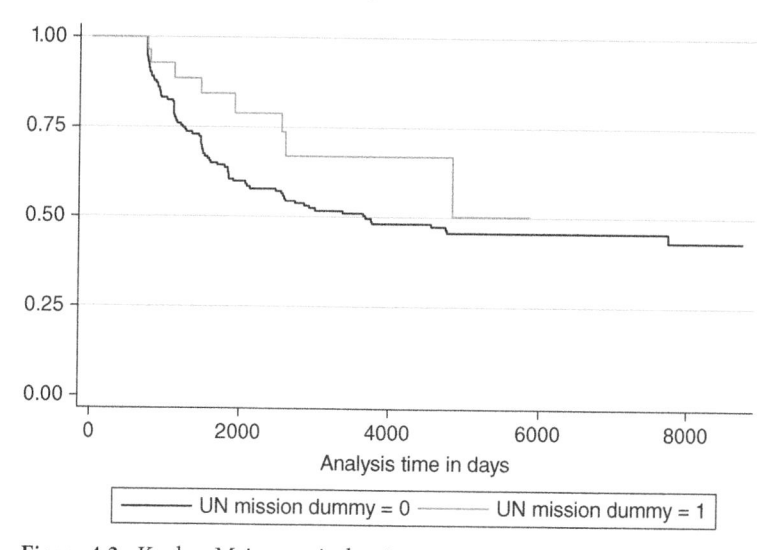

Figure 4.3. Kaplan-Meier survival estimates

Note: log-rank test for equality of survivor functions chi2(1) = 1.90, Pr > chi2 = 0.1682

dummy variable taking a value of 1 for the years during which the UNPKO is present. Although the line for peace spells with UNPKOs is above the line for those without, suggesting that UNPKOs are associated with longer peace spells, the formal test suggests that there is no significant difference between the spells with UNPKOs and those without. This is also the case when we only consider peace spells that lasted for a maximum of 4,000 days. We return to a discussion of UNPKOs and their contribution to peace durability below.

METHOD AND RESULTS

In our statistical analysis we want to examine which factors stabilize post-conflict peace. Applying survival analysis allows us to estimate what is known as a hazard function, which gives the probability that the event (end of peace) will occur given that the peace has lasted up to a specified time.[22] Our analysis uses a Cox proportional hazards

---

[22] We can write the hazard function as follows: $h(t) = h_0(t)\exp(x_j\beta_0)$ where $h_0(t)$ denotes the baseline hazard, the hazard common to all peace spells, j. The function exp () multiplies this baseline hazard, i.e., models how the explanatory variables, x,

88 *Measuring Peace*

model, a model that belongs to the category of semi-parametric models.[23] The use of the Cox proportional hazard model is popular in the study of the duration of peace; for example, it is used by both Fortna and Walter.

Before proceeding to our statistical model, we want to flag a number of potential problems. Our main aim is to explain peace stabilization and on the basis of our survival analysis we want to draw causal inferences. Ideally, we want our analysis to suggest that if some actions are taken, peace is more likely to endure. However, we have to be careful how we design and interpret our statistical analysis. When event A predates event B it is easier to justify the conclusion that A may cause B than in the situation when event A and B occur simultaneously. When event A and B occur simultaneously it could be that A causes B or that B causes A, or that an unknown event C drives both A and B. It is therefore important to consider simultaneity and endogeneity. In our case the characteristics of the conflict, such as fighting over territory and ethnic recruitment, happened before the event of peace. Similarly, the outcome of the conflict (victory, settlement, other) occurred before the event of peace. Thus, it is straightforward to include these variables in our model and to interpret them. On the other hand, income and peace are measured at the same time; they occur simultaneously. Peace is more likely to last if incomes are higher but incomes are also likely to be higher the longer the peace lasts, hence we have a problem of endogeneity. In order to guard against this endogeneity problem we can include lagged income, i.e., income that predates the event. The theoretical justification would be that past and current income are highly correlated.

The inclusion of UNPKOs in our model raises a similar problem. We observe UNPKOs and peace simultaneously. While UNPKOs may have an effect on the duration of peace it is also conceivable that the (expected) duration of peace has an effect on the decision to deploy a UNPKO and on the duration of the mission. The first issue is a problem of selection; if UNPKOs are predominantly sent to easier

---

shift the baseline hazard. The function exp () prevents the hazard $h(t)$ from taking negative values.

[23] Janet M. Box-Steffensmeier and Bradford S. Jones, *Event History Modeling: A Guide for Social Scientists* (New York: Cambridge University Press, 2004), ch. 4.

Factors of Post-Conflict Peace Stabilization 89

(harder) peace situations this would bias our results.[24] A positive (negative) coefficient would overestimate (underestimate) the impact of UNPKOs. Furthermore, the process that affects the changes in the UNPKO variable may be influenced by the duration of peace. Under this circumstance the usual interpretations of the explanatory variables in survival analysis do not hold. One solution would be to exclude such problematic variables. However, excluding explanatory variables that are theoretically relevant (notably UNPKO deployment) leads to model misspecification and potentially larger problems. From a policy advisory perspective, if we only use explanatory variables that are strictly exogenous, we will not be able to analyse a number of important policy issues. Hence, we simply flag these statistical problems and proceed with them in mind.

We can now develop a core model that enables us to investigate the impact of a number of key variables on the durability of peace. These key variables are: conflict outcome, characteristics of the armed conflict, and deployment of UNPKOs. As a starting point we present a model which only uses characteristics that occurred before the beginning of the peace spell: the outcome of the conflict, whether the conflict was fought over territory as opposed to governmental control, the duration of the conflict, and the intensity of the conflict (total number of battle deaths). This has two advantages: first it allows us to include all of the observations, and second these variables predate the peace spells and we do not have to worry about endogeneity and simultaneity issues. Rather than reporting coefficients, we report the hazard ratios. A hazard ratio greater than 1 suggests that this variable increases the hazard (or risk) of peace ending. The interpretation of hazard ratios is straightforward: a ratio of 1.5 suggests that a one unit change of the explanatory variable increases the hazard of the peace breaking down by 50 per cent (1–1.5 = –0.5). A hazard ratio of less than 1 suggests a decrease of the hazard ratio, i.e., making peace more durable. A hazard ratio of 0.4 suggests a 60 per cent reduction when the explanatory variable changes by one unit (1–0.4 = 0.6).

In our first model (Table 4.3, column 1) we include the dummy variables for the conflict outcome. Our category 'settlement' includes

---

[24] In her study of the initial post-Cold War period (1989–99), Fortna finds that UNPKOs are not deployed to the easiest cases (i.e., where conflicts have ended in a decisive outcome). See Fortna, 'Does Peacekeeping Keep Peace?'.

90　　　　　　　　　　*Measuring Peace*

peace agreements as well as ceasefires. The category 'other' includes cases of low or no activity as well as cases that do not meet other ACD criteria, e.g., one side ceased to exist. 'Victory' is the omitted category. The hazard ratios indicate that the hazard of a peace spell breaking down if the outcome is 'other' is 308 per cent higher than in the case of victory. Peace spells that ended with a settlement are 276 per cent more likely to break down than the comparison category, victory. Neither the duration of the conflict nor the intensity of the conflict (measured by the total number of battle deaths) are significant.

This first regression indicates that conflict termination is important for the likelihood of peace enduring and in the remainder of Table 4.3 we investigate this result in more detail. In the first model we classified both peace agreements and ceasefires as 'settlements' but in column 2 we investigate peace agreements and ceasefires separately. The results suggest that both ceasefires and peace agreements are more likely to break down than victories but that this hazard is greater for ceasefires. However, when we test for the equality of the

Table 4.3. Duration of peace and past conflict characteristics

	(1)	(2)	(3)	(4)
Outcome = other (low activity)	4.080***	4.138***	1.476**	1.470**
	(0.000)	(0.000)	(0.037)	(0.039)
Outcome = settlement	2.764**			
	(0.009)			
Outcome = peace agreement		2.074*		
		(0.076)		
Outcome = ceasefire		3.611***		
		(0.003)		
Outcome = victory			0.362***	
			(0.009)	
Outcome = government victory				0.234**
				(0.016)
Outcome = rebel victory				0.851**
				(0.743)
Conflict duration	0.999	0.999	0.999	0.999
	(0.705)	(0.634)	(0.705)	(0.807)
Conflict battle deaths	1.000	1.000	1.000	1.000
	(0.813)	(0.638)	(0.813)	(0.870)
Peace episodes	205	205	205	205
Number of observations	1925	1925	1925	1925
Number of failures	94	94	94	94

Note: hazard ratios reported, p-values in parentheses, dependent variable peace duration
* significant at 10 per cent; ** significant at 5 per cent; *** significant at 1 per cent

Factors of Post-Conflict Peace Stabilization 91

hazard ratios of peace agreements and ceasefires we can only reject this hypothesis at the 10 per cent level.[25] We then investigate the nature of the victory. First, we change the reference category from victory to settlement in Table 4.3, column 3. The results are the same as in column 1; however, changing the reference category means that we have to interpret the coefficient on the dummy variable victory as the inverse to the hazard ratio on settlement (1/2.76 = 0.36). In column 4 we include dummy variables for other, government victory, and rebel victory. The results suggest that although peace episodes are less likely to break down after government victories, they are not more likely to break down than after rebel victories. One reason, as Zeigler also suggests, may be that rebel movements are more prone to splintering.[26] However, we should keep in mind that there are only very few rebel victories (4 per cent of all terminations) which may account for the large standard error on the hazard ratio. When we test whether the hazard ratios for government and rebel victories are the same, we can only reject this hypothesis at the 10 per cent level.[27]

So far our results suggest that the severity of the armed conflict, measured as the duration of the conflict and the battle deaths caused, are not significant in the explanation of the duration of peace. In contrast, the termination of the armed conflict appears to be an important determinant of whether peace endures. Peace is much less likely to break down after military victories when compared to settlements[28] but these in turn are more likely to provide longer-lasting peace than in situations where the conflict activity was low but the conflict remained unresolved. When we investigate the nature of the victory or settlement we find some evidence that government victories are more stable than rebel victories and that peace agreements are followed by longer peace spells than ceasefires. However, the evidence is relatively weak and we continue our analysis without making distinctions within the categories 'settlement' and 'victory'.

In Table 4.4 we investigate the importance of a number of other explanatory variables. We start by including the dummy variable

---

[25] $\chi^2 = 2.84$, $p = 0.09$.

[26] Sean M. Zeigler, 'Competitive Alliances and Civil War Recurrence', *International Studies Quarterly* 60:1 (2016), 24–37.

[27] $\chi^2 = 2.86$, $p = 0.09$.

[28] We also know this from Licklider's work. See Roy A. Licklider, 'The Consequences of Negotiated Settlement in Civil Wars 1945–1993', *American Political Science Review* 89:3 (1995), 681–90.

Measuring Peace

Table 4.4. Duration of peace: territorial and ethnic conflicts and income

	(1)	(2)	(3)	(4)
Outcome = other	3.836***	1.659	3.374***	3.397***
	(0.001)	(0.316)	(0.004)	(0.004)
Outcome = settlement	2.631**	1.472	2.145**	2.256**
	(0.014)	(0.395)	(0.052)	(0.040)
Conflict duration	0.999	0.999	0.999	0.999
	(0.734)	(0.587)	(0.731)	(0.624)
Conflict battle deaths	1.000	0.999	0.999	0.999
	(0.786)	(0.567)	(0.575)	(0.799)
Territorial conflict	1.342			
	(0.169)			
Ethnic conflict		1.306		
		(0.430)		
Income (GDP) per capita			0.836*	
			(0.098)	
Peace episodes	205	131	178	178
Number of observations	1925	1385	1659	1659
Number of failures	94	47	77	77

Note: hazard ratios reported, p-values in parentheses, dependent variable peace duration
* significant at 10 per cent; ** significant at 5 per cent; *** significant at 1 per cent

territorial conflict. It takes a value of 1 if the conflict aim was territorial control and a value of 0 if the aim was government control. The hazard ratio for territorial conflict is not significant; however, including this variable violates the proportional hazards assumption.[29] In column 2 we add a dummy variable for ethnic armed conflict. The data are available from Wucherpfennig et al. and we code a conflict as ethnic if (1) the group makes a claim to operate on behalf of an ethnic group and (2) recruitment follows ethnic lines.[30] This variable is similar to the territorial conflict dummy: in 73 per cent of all the armed conflicts the conflict was ethnic and fought over territory or non-ethnic and fought over government control. The ethnic conflict dummy is insignificant and its inclusion violates the proportional hazards assumption.[31] Furthermore, the inclusion of ethnic conflicts changes the results considerably; no variable is significant. This is a model that not only violates the proportional hazards assumption but also has no explanatory value. The inclusion

[29] χ^2 = 12.47, p = 0.029.
[30] Julian Wucherpfennig, Nils W Metternich, Lars-Erik Cederman, and Kristian Skrede Gleditsch, 'Ethnicity, the State, and the Duration of Civil War', *World Politics* 64:1 (2012), 79–115.
[31] χ^2 = 31.09, p = 0.000.

Factors of Post-Conflict Peace Stabilization   93

of the ethnic war dummy reduces the sample size, instead of 205 peace episodes (corresponding to 1,925 observations) we can only consider 135 peace episodes (corresponding to 1,385 observations). In order to investigate the effect of sample size we re-estimate our core model of Table 4.3, column 1 and find that our results no longer hold on this reduced sample; it appears that the reduction in sample size affects the results significantly.

So far we have only considered information available from the ACD and from Wucherpfennig et al.; the latter reduced the number of observations considerably. Any concatenation with other data sets also causes a loss of observations. Often additional variables are not collected for some conflicts because the definition of conflict varies across data sets. Another reason is that data collection is difficult during armed conflict or in volatile situations. Thus, there are fewer economic variables available than political variables. Social scientists can determine that a country is at armed conflict (e.g., Somalia) but they are not able to collect data on population size, income, health, etc. Thus, one of the key questions is whether our empirical results remain intact when the sample size is reduced.

We turn to an examination of the effect of income in column 3. Income per capita is measured in purchasing power parity constant United States dollars, measured with a lag of two years, and we take the natural logarithm of this variable. Again, the inclusion of income reduces our sample size to 178 peace episodes (corresponding to 1,659 observations). Further investigation by running our core model on this reduced sample suggests that our main results still hold (column 4). Since our previous results hold on this reduced sample, we decide to include income per capita in our core model. Income has a positive effect on the duration of peace: societies with higher per capita income have a more lasting peace. The hazard ratio is significantly below 1, and an evaluation of the effect suggests that only large income changes are associated with a large reduction in the hazard of conflict recurrence. If a country with the minimum income ($142) increases its income to the average income ($3,605) the hazard decreases by 18.1 per cent. If a country increases its income from the average to the maximum income ($37,123) the hazard decreases by 7.9 per cent. Post-conflict economies often post high rates of economic growth owing to the low base period over which growth is measured. The fact that the average rate of economic growth in Burundi, one of our case studies, was only 4.1 per cent in the period

94 *Measuring Peace*

from 2004 to 2013 (compared with 7.4 per cent in Mozambique between 1993 and 2013; 9.8 per cent in Rwanda from 1995 to 2013, and 7.5 per cent in Sierra Leone between 2002 and 2013), may help to explain why the country is tottering on the brink of civil war as of this writing.[32]

We also investigated a number of other explanatory variables. None of the results were sufficiently strong to warrant inclusion in the core model. Remittances seem to have no effect on peace duration. There is possibly a small peace-enhancing effect from aid but donors may prefer to give aid to countries that appear to be more stable so the results may suffer from an endogeneity bias. We also investigated measures of vertical and horizontal inequality. (*Vertical inequality* consists in inequality among individuals or households; *horizontal inequality* is defined as inequality among groups.) However, this investigation is hampered by the number of missing observations. Our analysis suggests no effects from horizontal inequality and potentially a small beneficial effect from the reduction in vertical inequality. We find a small beneficial effect when we include the polity indicator to proxy for political regime. However, including the polity indicator is also problematic due to the fact that this composite indicator includes information about armed conflict.[33] Walter provides further analysis of governance indicators and suggests that the rule of law and public participation are important determinants in the survival of peace.[34] We also investigated whether peace spells in countries that grant regions autonomy last longer but unlike Collier et al. found no evidence.[35] We also found no evidence that elections have an impact on the hazard of peace ending. We considered as well the run-up to the election and the post-election year but found no evidence that the peace process is more likely to break down around election time.

In Table 4.5 we investigate the impact of UN peacekeeping operations (UNPKOs) led by the UN Department of Peacekeeping Operations. Qualitative data on the types of UNPKO are available from Howard and we updated these data for the purpose of this

---

[32] Janvier Nkurunziza's case study for this project.

[33] Scott Gates, Håvard Hegre, Mark Jones, and Håvard Strand, 'Institutional Inconsistency and Political Instability: Polity Duration, 1800–2000', *American Journal of Political Science*, 50:4 (2006), 893–908.

[34] Walter, 'Why Bad Governance Leads to Repeat Civil War'.

[35] Collier et al., 'Post-Conflict Risks'.

Factors of Post-Conflict Peace Stabilization   95

Table 4.5. Duration of peace and UNPKOs

	(1)	(2)	(3)	(4)	(5)	(6)
Outcome = other	3.406***	3.354***	3.438***	3.388***	3.328***	3.372***
	(0.004)	(0.005)	(0.004)	(0.004)	(0.004)	(0.004)
Settlement	2.341**	2.285**	2.320**	2.197**	2.238	2.671***
	(0.040)	(0.058)	(0.035)	(0.046)	(0.042)	(0.013)
Conflict duration	0.999	0.999	0.999	0.999	0.999	0.999
	(0.708)	(0.735)	(0.738)	(0.728)	(0.693)	(0.692)
Conflict battle deaths	0.999	0.999	0.999	0.999	0.999	0.999
	(0.611)	(0.694)	(0.572)	(0.566)	(0.562)	(0.581)
Income (GDP) per capita	0.834*	0.834*	0.806*	0.831*	0.811**	0.819*
	(0.105)	(0.102)	(0.061)	(0.094)	(0.064)	(0.082)
UNPKO (dummy)	0.583					
	(0.166)					
UNPKO (dummy during and after)		0.744				
		(0.372)				
UNPKO with DDR (dummy)			0.313**			
			(0.048)			
UN personnel				0.999		
				(0.475)		
Police					0.999***	
					(0.004)	
Observers					0.999	
					(0.303)	
Troops					1.000**	
					(0.035)	
Settlement*UNPKO						0.209**
						(0.017)
Peace episodes	178	178	178	178	178	178
Number of observations	1659	1659	1659	1659	1659	1659
Number of failures	77	77	77	77	77	77

Note: hazard ratios reported, p-values in parentheses, dependent variable peace duration
* significant at 10 per cent; ** significant at 5 per cent; *** significant at 1 per cent

study (Table 4.6).[36] Quantitative data on UNPKOs are available from the International Peace Institute data base, which provides information on UN personnel: how many troops, police officers, and observers were present and who the contributing countries were.[37] We begin by simply including a dummy variable indicating the presence of a UNPKO (column 1). The hazard ratio indicates that UNPKOs decrease the hazard of the peace ending but the hazard ratio is not

[36] Lise Morjé Howard, *UN Peacekeeping in Civil Wars* (Cambridge: Cambridge University Press, 2008), updated by Kate Roll.
[37] International Peace Institute, *IPI Peacekeeping Database*, www.providingforpeace-keeping.org.

96　　　*Measuring Peace*

Table 4.6. UN peacekeeping operations

Acronym	Peacekeeping operation name	Start date	End date
UNIFIL	United Nations Interim Force in Lebanon	March 1978	Present
ONUCA	United Nations Observer Group in Central America	November 1989	January 1992
UNAVEM II	United Nations Angola Verification Mission II	June 1991	February 1995
ONUSAL	United Nations Observer Mission in El Salvador	July 1991	April 1995
UNPROFOR	United Nations Protection Force	February 1992	March 1995
ONUMOZ	United Nations Operation in Mozambique	December 1992	December 1994
UNOMIG	United Nations Observer Mission in Georgia	August 1993	June 2009
UNOMIL	United Nations Observer Mission in Liberia	September 1993	September 1997
UNMIH	United Nations Mission in Haiti	September 1993	June 1996
UNAMIR	United Nations Assistance Mission for Rwanda	October 1993	March 1996
UNMOT	United Nations Mission of Observers in Tajikistan	December 1994	May 2000
UNAVEM III	United Nations Angola Verification Mission III	February 1995	June 1997
UNCRO	United Nations Confidence Restoration Operation in Croatia	May 1995	January 1996
UNMIBH	United Nations Mission in Bosnia and Herzegovina	December 1995	December 2002
UNTAES	United Nations Transitional Administration for Eastern Slavonia, Baranja and Western Sirmium	January 1996	January 1998
UNMOP	United Nations Mission of Observers in Prevlaka	January 1996	December 2002
UNSMIH	United Nations Support Mission in Haiti	July 1996	July 1997
MINUGUA	United Nations Verification Mission in Guatemala	January 1997	May 1997
MONUA	United Nations Observer Mission in Angola	June 1997	February 1999
UNTMIH	United Nations Transition Mission in Haiti	August 1997	December 1997
MIPONUH	United Nations Civilian Police Mission in Haiti	December 1997	March 2000
UNCPSG	United Nations Civilian Police Support Group	January 1998	October 1998
UNOMSIL	United Nations Observer Mission in Sierra Leone	July 1998	October 1999
UNMIK	United Nations Interim Administration Mission in Kosovo	June 1999	April 2015

UNTAET	United Nations Transitional Administration in East Timor	October 1999	May 2002
UNAMSIL	United Nations Mission in Sierra Leone	October 1999	December 2005
UNMISET	United Nations Mission of Support in East Timor	May 2002	May 2005
UNMIL	United Nations Mission in Liberia	October 2003	Present
UNOCI	United Nations Operation in Côte D'Ivoire	April 2004	Present
ONUB	United Nations Operation in Burundi	June 2004	December 2006
MINUSTAH	United Nations Stabilization Mission in Haiti	June 2004	Present
UNMIT	United Nations Integrated Mission in Timor-Leste	August 2006	December 2012
UNISFA	United Nations Interim Security Force for Abyei	June 2011	Present
MINUSMA	United Nations Multidimensional Integrated Stabilization Mission in Mali	April 2013	Present

Note: 'present' marks the end of the period of observation, which ended on 31 December 2013. We only list United Nations peacekeeping operations for post-conflict periods that we could include in our analysis presented in Table 4.5

significant at conventional levels (p = 0.16). We proceed by investigating whether UNPKOs have an 'innoculation effect', i.e., we include a dummy taking a value of 1 while the operation was in place and for all subsequent years (column 2). UNPKOs do not 'innoculate' against conflict recurrence: there is no statistically significant difference between the duration of peace spells with and without UNPKOs.

On the basis of these two models we investigate whether the type of UNPKO matters. In column 3 we include a dummy for missions that had a mandate for the disarmament, demobilization, and reintegration (DDR) of armed forces. We find that these missions significantly lower the hazard of the peace breaking down: they decrease the hazard by 69 per cent. We further tried dummies for UNPKOs that had troops on the ground, i.e., excluding operations with police and/or observers only. We also constructed a dummy for peace enforcement operations and a dummy variable for UNPKOs that were not confined to their base. None of these variables were statistically significant.

We then turn to the analysis of the effect of UN personnel. In column 4 we simply include the number of UN personnel; this figure comprises troops, police, and observers. This variable is

98 *Measuring Peace*

insignificant. In column 5 we investigate the effect of troops, police, and observers separately. The results indicate that observers appear to have no effect on the hazard of peace breaking down: troops increase the hazard and police lower it. Evaluating the change in the hazard by comparing no troops with the average number of troops (5,340) we find that the hazard increases by 48 per cent. When police forces are increased from zero to the mean (790) the hazard decreases by 43 per cent.

In the last column of Table 4.5 we include an interaction term of peace settlements and UNPKOs.[38] The hazard ratio is less than 1, indicating that the deployment of UNPKOs supports peace settlements. The effect is large, for peace settlements without UNPKOs the hazard of peace ending is 167 per cent higher but for peace settlements that are supported by UNPKOs the hazard of peace ending is about 44 per cent lower. Even though this is an interesting result, it rests on a relatively small number of observations. Only 34 out of 205 peace episodes had a UNPKO, of which 20 were deployed after settlements.[39] We return to this finding below.

There were a number of other variables that we tried but found no statistical significance for. Economic variables included economic growth, development aid, and remittances. Political indicators included the polity indicator from the Polity IV data and elections.

There were also a number of factors that our case study authors considered important for their role in sustaining the peace which we found too difficult to measure or for which we lack comprehensive data. These included strategic conditions (e.g., stalemate), national leadership qualities, elite political cooperation and cohesion among parties to the conflict, the behaviour of regional actors, the use of transitional justice mechanisms, and inclusive settlements/governance.

---

[38] Cleves et al. provide a guide to the interpretation of interaction terms. Mario Cleves, Roberto G. Gutierrez, William Gould, and Yulia V. Marchenko, *An Introduction to Survival Analysis Using Stata*, 3rd edn (College Station, TX: Stata Press, 2010), 186–9. They stress that the inclusion of interaction terms does not necessitate the inclusion of the corresponding main effects. The shift of the baseline hazard is calculated in the following way: the coefficient estimates are simply the natural logarithms of the hazard ratios. For settlement the coefficient is $\ln(2.6714) = 0.9826$ and for the interaction term UNPKO*settlement the coefficient is $\ln(0.2091) = -1.5651$. The hazard ratio for observations that experienced a settlement and a UNPKO is thus $\exp(0.9826 - 1.5651) = 0.558$.

[39] In total there were thirty-three peace episodes that received UNPKOs at some stage: twenty after settlements, six after victories, and seven in situations of 'other'.

Factors of Post-Conflict Peace Stabilization 99

Some of these factors have been examined in the literature, including a few studies that employ survival analysis.[40] There were also a number of variables emerging from the case studies that undermined or threatened to undermine the peace, notably corruption/bad governance, impunity, elite political rivalries, lack of inclusiveness, unresolved property disputes, and youth unemployment. These factors also bear further systematic consideration.

DISCUSSION

As we noted at the outset of this chapter, it is difficult to explain the duration of peace. However, in our regressions we established a number of empirical regularities. One robust statistical result is that victories provide more long-lasting peace than settlements and that unresolved conflicts (measured by category 'other') are most likely to break down. There is some evidence that peace agreements provide a longer-lasting peace than ceasefires and that in cases of government victory the peace lasts longer than in cases of rebel victory.

We find no evidence that peace duration after territorial or ethnic conflicts is different from conflicts over governmental control or that the severity of the armed conflict, measured as conflict duration or battle deaths, has an impact on the duration of peace. Ethnic conflicts tend to last longer. Wucherpfennig et al. argue that ethnic exclusionary policies make it less likely for governments to accept settlements and rebel groups tend to have stronger group solidarity and are thus able to fight for longer.[41] However, we find that the length of conflict

---

[40] On rebel group competition/fragmentation and its impact on peace duration, see Zeigler, 'Competitive Alliances and Civil War Recurrence' and Rudloff and Findley, 'The Downstream Effects of Combatant Fragmentation on Civil War Recurrence'; on other organizational characteristics of rebel groups, see John Ishiyama and Anna Batta, 'Rebel Organizations and Conflict Management in Post-Conflict Societies 1990–2009', *Civil Wars* 13:4 (2011), 437–57; on features of power-sharing arrangements and their impact on peace duration, see Remzi Badran, 'Intrastate Peace Agreements and the Durability of Peace', *Conflict Management and Peace Science* 31:2 (2014), 193–217; Martin, 'Coming Together'; and Melani Cammett and Edmund Malesky, 'Power Sharing in Postconflict Societies: Implications for Peace and Governance', *Journal of Conflict Resolution* 56:6 (2012), 982–1016; on inclusive peace settlements, see Call, *Why Peace Fails*.

[41] Wucherpfennig et al., 'Ethnicity, the State, and the Duration of Civil War'.

has no significant impact on peace duration. On the other hand, a smaller proportion of ethnic conflicts end in settlement (35 per cent for ethnic conflicts as opposed to 43 per cent for all conflicts) and a higher proportion of ethnic conflicts rumble on below the ACD threshold (46 per cent for ethnic conflicts versus 40 per cent for all conflicts).[42]

We also examined indicators of horizontal and vertical inequality. We find no evidence that measures of horizontal inequality have an impact on the duration of peace but find some evidence that vertical inequality has a negative impact on the duration of peace. However, the sample size was greatly reduced by the inclusion of any inequality measure and these results should be treated with caution.

For UN peacekeeping we find little evidence that the presence of UNPKOs has a stabilizing effect on peace. This is in contrast to Fortna, who finds a positive effect of UNPKOs on the duration of peace.[43] One of the reasons why her results are different may be due to the fact that she uses a different data source for the definition of peace (based on Doyle and Sambanis) and that her sample only covers 1990–9.[44] She herself points out that her sample size is small, thus her results should be interpreted with caution. However, our results tally with Walter.[45] She uses the same data source to define peace spells (ACD) and applies the method of Cox proportional hazard regressions. Like us she finds no evidence that UNPKOs stabilize the peace.

However, we do find some evidence that UNPKOs with a DDR component enhance the peace. We also find evidence that the presence of police forces in the mission contributes to peace duration. And, finally, we find that UNPKOs have a positive effect on peace duration when the conflict ends in a settlement. Due to the small number of observations we cannot tell whether this effect is stronger after peace agreements than after ceasefires.

One possible explanation for the peace-stabilizing effect of a UNPKO after a settlement could be that the UN was instrumental in settling the conflict. In our study we restrict our analysis to the post-conflict period but most UNPKOs were deployed before the

---

[42] Own calculations.   [43] Fortna, 'Does Peacekeeping Keep Peace?'
[44] Michael Doyle and Nicholas Sambanis, *Making War and Building Peace* (Princeton, NJ: Princeton University Press, 2006).
[45] Walter, 'Why Bad Governance Leads to Repeat Civil War'.

Factors of Post-Conflict Peace Stabilization   101

armed conflict ended. Out of the thirty-three UnPKOs that we include in our statistical analysis, twenty started before the end of the armed conflict as coded in the Armed Conflict Dataset. The research by Hegre et al. examines the likelihood of transitions between peace, minor conflict, and major conflict.[46] Their results suggest that UNPKOs have a stabilizing effect. The main pathway appears to be through depressing violence during conflict: minor conflicts do not scale up into major conflicts but through the presence of a UNPKO the transition from minor conflicts to peace becomes more likely. This indicates that UNPKOs may be less about 'keeping' the peace than 'preparing' for peace—an effect that we cannot study in our survival analysis. However, El Salvador, one of our case studies, provides some evidence in support of this observation. In El Salvador, where there has been no recurrence of civil war, the UN deployed observers in support of a human rights agreement and before a ceasefire was in place.[47]

In order to make this statistical result meaningful it is instructive to consider the case studies as to why UNPKOs make the peace last longer. Five of the six cases examined for this study were host to a UNPKO of varying size, duration, and mandate (see Table 4.7); Nepal was a special political mission not led by the UN Department of Peacekeeping Operations. All of the operations were deployed in support of a peace agreement. In El Salvador, the UN mission (ONUSAL) played a key role keeping implementation of the 1992 peace agreement on track, notably with regard to demobilization and demilitarization, arms control, and human rights verification. In the case of Burundi, a peacekeeping force was deployed in 2003 after the conclusion of the Arusha Agreement. Without foreign troops (first African Union forces and then UN peacekeepers (ONUB)) to protect Burundian politicians who came back from exile, it is doubtful that Burundi would have experienced the political transition which ended the forty-year rule by a minority of elites (although at the time of writing that peace is in jeopardy again). In Liberia, the UN mission (UNMIL) provided a crucial security guarantee that assured civil society the safety it needed after the 2003 Accra Accord to participate

---

[46] Håvard Hegre, Lisa Hultman, and Håvard Mokleiv Nygård, 'Evaluating the Conflict-Reducing Effect of UN Peacekeeping Operations', mimeo (2014), https://www.dropbox.com/s/m1k612fg8vg1syc/PKO_prediction_2013.pdf.
[47] Charles T. Call's case study for this project.

102   *Measuring Peace*

Table 4.7. UNPKOs and peace settlements (case studies)

Country	UNPKO start	Peace settlement	Peace start
Burundi	June 2004: ONUB deployed (previously South African force: from September 2001)	August 2000: Arusha Agreements; November 2003: accord signed by FDD	November 2003
East Timor	February 2000: UNTAET deployed (previously: September 1999: INTERFET, a non-UN force)	May 1999 Agreement between Indonesia and Portugal	September 1999
El Salvador	July 1991: ONUSAL deployed	July 1990: Human rights agreement; January 1992: final accord	January 1992
Liberia (2nd civil war)	October 2003: UNMIL (previously ECOMIL)	August 2003: Accra Accord	July 2003
Sierra Leone	October 1999: UNAMSIL	January 1990: Lomé Peace Accord	January 2002

effectively in political life. And in East Timor, the UN-authorized, Australian-led international force (INTERFET) helped to stabilize the territory following the violence wrought by Indonesian-backed militia. (Subsequent UNPKOs were important for the pursuit of serious crimes and the creation of order during the transitional period in the absence of national police and military.) Counterfactual reasoning adds further support for the contribution of UNPKOs to stabilization: without peacekeepers, it is easy to imagine that East Timor, Liberia, and Burundi could have returned to war.[48]

CONCLUSIONS

Our survival analysis of the duration of post-conflict peace suggests that it is difficult to identify determinants of peace stability. A number of conflict-specific variables are not statistically significant, e.g., measures of the severity of the conflict (armed conflict duration and number of battle deaths). Conflicts are fought over government or

[48] We owe this latter observation to Alex Bellamy.

Factors of Post-Conflict Peace Stabilization   103

territorial control, but whether the fighting is over territorial control or to take over government does not appear to have an impact on the duration of the peace. However, there is some indication that the type of conflict termination is a predictor of the stability of the peace. Military victories, in particular by the government, make the peace last longer. Income appears to stabilize the peace but there are the usual concerns regarding endogeneity and simultaneity, even though we lag per capita income. Other economic variables, such as growth, aid, and remittances, were not found to be statistically significant. Our investigation of vertical and horizontal inequality also suggests that these variables are not statistically significant.

We also examined the impact of UNPKOs. There is some previous work suggesting that UNPKOs in their own right stabilize the peace (Fortna 2004 and Collier et al. 2008) but we found no such evidence. This may be due to different definitions of conflict (we use ACD data) or the larger number of observations that we have. In any case, we find some evidence that settlements are made more stable by UNPKOs. However, we have to keep in mind that the sample size is relatively small and that the results are sensitive to small changes in sample size. This is not uncommon when using cross-country data.

Why might UNPKOs matter in relation to a political settlement? One reason is that a UNPKO can raise the profile of a conflict-affected country, generating greater regional/international interest in and support for peacebuilding there. Much also depends on the precise role a UNPKO performs, which will vary from case to case. UN forces can play an important role in the verification of arms and other agreements, in fostering conditions conducive to the holding of elections, and in creating a secure environment for civil society to engage, among other positive contributions. In order to find out more about the relationship between UNPKOs and their stabilizing role in post-conflict situations after settlement it is instructive to look at our country case studies. Five of the six cases involved the deployment of a UNPKO after a settlement. In each case it is possible to identify specific contributions that the peacekeeping operation contributed to peace stabilization. As there are only twenty peace episodes that see UNPKOs deployed after a settlement, it would be possible to conduct a more focused examination of all of them to establish the nature and the extent of any causal links. This is left for future research.

5

Measuring Peace Consolidation

> [Peace] becomes sustainable, not when all conflicts are removed from society, but when the natural conflicts of society can be resolved peacefully.
>
> United Nations[1]

We return now to the primary question that animates this study: can we know whether the peace that has been established in the aftermath of an armed conflict is a stable peace and, if so, how can we know?

This chapter will discuss the implications of the analysis in the foregoing chapters for efforts—actual or potential—to ascertain the robustness of post-conflict peace. What does this analysis tell us about whether and how we can determine the strength or vulnerability of a peace? How can monitoring and assessment of peace consolidation be enhanced? Are there 'natural' limits to what can be known about the quality of the peace in a given post-conflict environment? This chapter will also discuss the obstacles to sound analysis and good practice and how some of these obstacles might be overcome.

FOUNDATIONAL MATTERS

To measure the robustness of a peace, and progress towards achieving it, it is necessary to define peace: one cannot measure what has not been defined. As we observed in Chapters 1 and 2, peace as both an

---

[1] United Nations, 'No Exit without Strategy: Security Council Decision-Making and the Closure or Transition of United Nations Peacekeeping Operations', Report of the Secretary-General, 20 April 2001, UN Doc. S/2001/394, para. 10.

Measuring Peace Consolidation

object of academic study and as a policy goal can have many different meanings. At the very least, peace can mean the non-recurrence of armed conflict; more ambitiously, peace can mean the elimination of direct personal violence and indirect structural violence. To be usable, however, these simple definitions require further elaboration: what is the threshold of violence beyond which peace can be said to have failed? How much structural violence constitutes a failed peace? Moreover, who is to make these determinations and on what basis?

From a scholar's standpoint, many of these distinctions are matters of academic convention that enjoy consensus—albeit sometimes a contested consensus—for the sake of facilitating analysis across cases and studies. Without common terms of reference (e.g., with regard to what constitutes civil war) there can be little or no basis of comparison and therefore no meaningful dialogue or debate. Indeed, it is often the lack of agreement on key terms of reference that accounts for divergent assessments among scholars.[2] From the practitioner's standpoint, clarity can be even more elusive. The reason for this is that while a significant body of scholarship—quantitative analysis in particular—employs a minimalist 'negative' notion of peace, most practitioners, as we saw in Chapter 2, adopt broader conceptions of peace that are inherently less precise. However, the imprecision is not necessarily insuperable; to a certain degree, as we will see below, it can be overcome.

Another way to approach the question of what defines a peace is to ask, what are the characteristics of and requirements for a stable peace? There is both an empirical and a normative dimension to this question. It is an empirical question in much the same way as one might investigate how much weight a bridge can support or what the melting point of a given material is. The analogy is apt because peace, like bridges or materials, can collapse or melt when subjected to too much stress of a particular kind.[3] The difference is that one can know

---

[2] Compare, for instance, Roland Paris's positive assessment of peacebuilding in Eastern Slavonia/Croatia with Peter Wallensteen's comparatively negative assessment, which turns on the salience of 'dignity' as a requisite factor. Roland Paris, *At War's End: Building Peace after Civil Conflict* (Cambridge: Cambridge University Press, 2004), 107–10; Peter Wallensteen, *Quality Peace: Peacebuilding, Victory and World Order* (New York: Oxford University Press, 2015), 39–41.

[3] However, what are the implications of a 'partial' or 'temporary' collapse of a peace? For 'conventional' quantitative analysts, the thresholds discussed in Chapter 4 apply. For other analysts it is a matter of judgement (about which more below).

106   *Measuring Peace*

the physical properties of a given material and one can know with a high degree of confidence how that material will behave under certain conditions. The same cannot be said of a peace—or at least not with the same degree of confidence—because its properties are often more opaque owing to the complexity of the conflict dynamics that underlie a given conflict situation. And there may be external shocks—regional conflict, refugee flows, natural disaster—that will take their toll and cannot always be anticipated or controlled for. This is not to suggest the total lack of a relevant knowledge base; that past experiences—gleaned from comparative case studies or statistical analysis—can offer no insights. Rather, they can point towards avenues of potentially fruitful investigation, as we observed in Chapter 4 with regard to peace stabilization. However, they need to be supplemented by methods and approaches that take into account the particularities of a given conflict situation. As Paul Collier and associates have acknowledged with respect to their own research:

> [A]ny statistical analysis such as our own needs to be supplemented by appropriate contextual knowledge before being applied in any particular situation. There are limits to how much the past can be a guide to the future and there are even stricter limits to how much a statistical analysis can extract from this past experience.[4]

We return to the matter of 'appropriate contextual knowledge' below.

There is not only an empirical dimension to how robust a peace may be but also a normative dimension. Peace is elastic (although not infinitely so): the characteristics of peace can vary significantly and yet remain 'faithful' to the concept. Peace in one society may look different to peace in another society. One society may tolerate apparent threats to or breaches of the peace that another society would find unacceptable. Peace is constructed in part on the basis of the beliefs, opinions, and preferences of those directly involved—the conflict-affected society. One society may choose, in the interest of peace, to prosecute atrocities committed during violent conflict; another society may choose, also in the interest of peace, to eschew accountability, bearing in mind that societies will often be divided on this and other

---

[4] Paul Collier, Anke Hoeffler, and Måns Söderbom, 'Post-Conflict Risks', *Journal of Peace Research* 45:4 (2008), 474.

Measuring Peace Consolidation 107

fundamental questions of peacebuilding.[5] Whether or not a society is at peace, then, needs to take account of the beliefs, opinions, and preferences of those directly involved, which, to a large extent will be the outgrowth of local historical experience and local value systems.[6] Of course those directly affected may be unaware of how strong or fragile their peace actually is.[7] This is where the vast body of scholarship—notwithstanding its inconclusiveness and inconsistencies—can be useful in generating guidance or producing indicators and other measures of peace. However, the bottom line is that any sound assessment of a peace must be built on knowledge that reflects the local context. The question is not 'what does peace require?' but 'what does *this* peace require?'

This may seem obvious. However, peacebuilding organizations have generally failed to adopt modes of operation that facilitate the acquisition of knowledge that is vitally important for assessing the quality of the peace that prevails in a given post-conflict environment. In the section that follows, I draw on scholarship and practitioner analysis to diagnose the nature of this chronic problem and to identify the principles that would need to underpin assessment practices to make them more effective.

AN ETHNOGRAPHIC APPROACH

There has been a burgeoning of critical peacebuilding studies in the past decade, among which scholarship figures prominently that of Séverine Autesserre. In her now classic study of international peacebuilding, *Peaceland*, Autesserre shows how 'standard modes of operation' of international peacebuilding organizations militate against

---

[5] On transitional justice in relation to peacebuilding, see Chandra Lekha Sriram, 'Justice as Peace? Liberal Peacebuilding and Strategies of Transitional Justice', *Global Society* 21:4 (2007), 579–91.

[6] Another way to express this is to say that the question of what is peace is bound up with normative questions about the quality of life a given community accepts for itself.

[7] The tension between 'subjectivism' and 'objectivism', as discussed by Chris Mitchell in relation to conflict, has clear parallels to peace. See Chris Mitchell, 'Recognising Conflict', in Tom Woodhouse (ed.), *Peacemaking in a Troubled World* (New York: Berg, 1991), 209–25.

108   *Measuring Peace*

peacebuilding effectiveness. She identifies several factors that explain why peacebuilding often fails to reach its full potential. These include a tendency within peacebuilding organizations to favour broad 'thematic expertise over local knowledge'; a preference for 'technical, short-term, and top-down solutions to complex social, political, and economic problems', an orientation towards analysis that emphasizes 'quantifiable results', the compartmentalization and 'stove-piping' of peacebuilding activities, and personal and social practices among international personnel that 'create boundaries between them and host populations' which, in turn, militate against the transmission of local knowledge.[8]

While Autesserre's observations are directed towards the practices of international peacebuilders broadly, many of these observations have relevance to measuring peace consolidation specifically. To begin with, her method of research—what she calls an 'ethnographic approach'—deserves consideration as a first principle of sound analysis. Although conceived with international peacebuilding organizations in mind as the object of her research, the approach can also be used to characterize a more constructive means of conducting 'strategic assessment', which is the primary concern of this volume. An ethnographic approach, for the purposes of strategic assessment, would favour greater reliance on knowledge of local culture, local history, and especially, the particular conflict dynamics at work in a given conflict, including at the micro level. 'Familiarity with local context is crucial,' Autesserre observes. 'In-depth knowledge of local histories, cultures, traditions, attitudes, and worldviews . . . remains critical throughout project implementation.'[9]

A good example of an ethnographic approach to conflict analysis/ strategic assessment can be found in the work of a small number of analysts in relation to East Timor (Timor-Leste), where the United Nations (UN) authorized the deployment of an intervention force (INTERFET) to restore peace and security in September 1999 and then, from 2000, established a series of peacebuilding operations to

[8] Séverine Autesserre, *Peaceland: Conflict Resolution and the Everyday Politics of International Intervention* (Cambridge: Cambridge University Press, 2014), 249.

[9] Ibid., 70. Two exemplary studies of conflict that reflect an ethnographic approach are David Kilcullen, *The Accidental Guerrilla: Fighting Small Wars in the Midst of a Big One* (London: C. Hurst & Co., 2017) and Jeffrey Race, *War Comes to Long An: Revolutionary Conflict in a Vietnamese Province*, 2nd edn (Berkeley, CA: University of California Press, 2010).

Measuring Peace Consolidation 109

administer the territory pending its independence in 2002 and to support the consolidation of peace following independence.[10] UN efforts in East Timor were viewed widely as a success in terms of establishing and maintaining a secure environment and facilitating the construction of nascent state and governmental institutions—at least until the outbreak of renewed violence in April 2006.[11] The recurrence of violence, which left scores dead and displaced some 150,000 others (out of a total population of 1 million), caught most international observers unawares, including many of the practitioners engaged in peacebuilding in East Timor.

The violence came as a surprise to many because unrest previously had originated with neighbouring Indonesia and pro-Indonesian forces which had opposed East Timor's independence. Now that East Timor was independent, had its own national defence and police forces, and enjoyed the support of the international community, these elements no longer appeared to pose a major threat. The trouble is that the dangers to security were also internal. Edward Rees, a UN official serving in East Timor, warned in 2004 of the growing politicization of the national defence and police forces and the absence of adequate civilian oversight and management mechanisms which had already led to clashes between these forces in 2002 and 2003.[12] The United States Agency for International Development (USAID) warned, too, in a 2004 Conflict Vulnerability Assessment, that 'localized, violent outbursts/conflicts' were 'inevitable' and identified presciently the key actors and sectors of the population who were likely to

---

[10] For an overview of UN engagement in peacebuilding in East Timor, see Katsumi Ishizuka, *The History of Peace-Building in East Timor: The Issues of International Intervention* (Cambridge: Cambridge University Press, 2010).

[11] As the UN Secretary-General wrote in his report to the Security Council of 26 July 2000: 'Today, although it has not yet reached its full designated capacity, UNTAET [the UN transitional administration in East Timor] can look with satisfaction on what it has achieved so far. It has contributed to the alleviation of the emergency brought about by the violence and destruction that followed the popular consultation last year; it has maintained a secure environment; it has established the foundations of an effective administration; and, above all, it has established a relationship of mutual respect and trust with the East Timorese.' 'Report of the Secretary-General on the United Nations Transitional Administration in East Timor', UN Doc. S/2000/738, 26 July 2000, para. 63.

[12] Edward Rees, 'Under Pressure. FALINTIL–Forças De Defesa De Timor-Leste: Three Decades of Defence Force Development in Timor-Leste 1975–2004'. Working Paper, no. 139. Geneva Centre for the Democratic Control of Armed Forces, April 2004.

110   *Measuring Peace*

become involved in conflict activities.[13] James Scambary, an independent researcher based in the capital city of Dili, also became aware
of the deteriorating security situation in East Timor at the time as a
consequence of his extensive engagement with local Timorese. He
would later produce a comprehensive anatomy of the violence that
took account of 'multiple fracture lines' and 'social tensions' in East
Timor which were scarcely evident to the many international peacebuilders working alongside him.[14] What these and a handful of other
analysts had in common was strong contextual knowledge, first and
foremost, which afforded them insights into local conflict-relevant
dynamics and their implications for the maintenance of peace. Of
course, it can sometimes be difficult to ascertain the reliability of pre-
conflict analysis, especially when there is a lot of dissonant 'noise', but
in this case expert opinion (among these experts at least) was pointing
more or less in the same direction. The real problem is that by 2006
there was a strong imperative from donors and troop-contributing
countries to redeploy resources from East Timor that made these
actors 'tone deaf' to appeals from well-informed analysts.

Looking back on US misadventures in Vietnam, for which he bore
considerable responsibility, former US secretary of defense Robert
McNamara would bemoan 'our profound ignorance of the history,
culture, and politics of the people in the area'.[15] Similarly, the British
MP Rory Stewart, writing forty years after McNamara, would identify
the lack of local knowledge on the part of international personnel as 'a
central cause of the humiliating mess in Afghanistan' (and elsewhere), where Stewart had served in peace- and state-building operations.[16] Indeed, nearly eight years after intervening in Afghanistan,
neither the United Kingdom (UK) Foreign Office nor the UK Department for International Development had a single Pashto speaker on

---

[13] United States Agency for International Development, 'The Crisis in Timor-
Leste: Causes, Consequences, and Options for Conflict Management and Mitigation',
November 2006, https://apcss.org/core/Library/CSS/CCM/Exercise%201/Timor%
20Leste/2006%20Crisis/USAID%20Conflict%20Assessment%20Nov%202006.pdf. See
annex 1 for a summary of the 2004 conflict vulnerability assessment's conclusions
and recommendations.

[14] James Scambary, 'Anatomy of a Conflict: The 2006–2007 Communal Violence
in East Timor', *Conflict, Security and Development* 9:2 (2009), 265–88.

[15] Robert McNamara, *In Retrospect: The Tragedy and Lessons of Vietnam*
(New York: Times Books, 1995), 322.

[16] Rory Stewart and Gerald Knaus, *Can Intervention Work?* (New York: W. W. Norton &
Co., 2011), 13.

Measuring Peace Consolidation 111

the ground, even though Pashto was the language of Helmand province where the UK had principal responsibility for the Provincial Reconstruction Team.[17] Theo Farrell, too, in his analysis of Britain's war in Afghanistan from 2001, highlights the insufficient grasp of local knowledge, notably the consistent failure to 'analyze the underlying political dynamics of the conflict, and the constraints imposed by host nation and regional politics', as one among several factors that contributed to the UK's (and US's) inability to achieve its objectives.[18]

What we often see instead of an ethnographic approach to peace and conflict is a generic approach that is underpinned by generalizable or 'thematic' knowledge. As both Autesserre and Stewart observe, foreign service personnel are trained increasingly in such generalizable skills as project management, public finance, agricultural engineering, gender mainstreaming, donor coordination, and the like. At the same time, in the US and elsewhere, government support for area and regional studies has been declining, thus further impoverishing this valuable knowledge base.[19] As a result, Stewart observes of British civilians serving in Afghanistan, 'their knowledge about Afghanistan itself—or indeed about any other developing country—[is] generally much more limited than that of a previous generation of foreign service officers'.[20] While generalizable expertise is extremely valuable, indeed indispensable, the emphasis placed on it tends to reduce peace consolidation to a technical problem of building capacity, thus failing to take into account sufficiently the fundamentally political character and context of the process of peacebuilding.[21]

---

[17] UK House of Commons, Foreign Affairs Committee, 'Eighth Report, Global Security: Afghanistan and Pakistan, Session 2008–9' (2009), para. 250, https://publications.parliament.uk/pa/cm200809/cmselect/cmfaff/302/30209.htm.

[18] Theo Farrell, *Unwinnable: Britain's War in Afghanistan 2001–2014* (London: Bodley Head, 2017), 421. The US military made an effort to overcome this deficiency with its Human Terrain programme, which sought to bring anthropologists and other social scientists with regional-specific knowledge of Afghanistan (and Iraq) and linguistic skills to the battlefield, deployed alongside US soldiers, to provide them with advice about local power structures and conflict dynamics. However, the programme failed to recruit sufficient numbers of suitable specialists.

[19] Charles King, 'The Decline of International Studies', *Foreign Affairs* 94:4 (July/August 2015). See also British Academy, *Lost for Words: The Need for Languages in UK Diplomacy and Security* (London: British Academy, 2013).

[20] Stewart and Knaus, *Can Intervention Work?*, 17.

[21] Mats Berdal and Dominic Zaum, 'Power after Peace', in Mats Berdal and Dominic Zaum (eds), *The Political Economy of Statebuilding: Power after Peace* (Abingdon: Routledge, 2013).

112   *Measuring Peace*

Yet, to succeed in peacebuilding generally and, one can add, in strategic assessment specifically, it is important, as Westendorf points out, to 'engage with the politics of conflict and peace in the postwar society, particularly in terms of how power and authority are organized and contested, and how competing interests intersect with either peace building or the continuation of conflict'.[22] Peacebuilders are often woefully ill-equipped to comprehend this.

Complementary to the knowledge that an ethnographic approach can produce are two other related sources of local knowledge that can be valuable for assessing the robustness of the peace: knowledge derived from forensics and knowledge derived from intelligence. The violence that broke out in UN-administered Kosovo in March 2004 provides evidence of the utility of both. Formerly an autonomous province of Serbia within Yugoslavia, Kosovo came under UN administration following a North Atlantic Treaty Organization (NATO) military campaign in 1999 against Serbia whose repression of the majority ethnic Albanian community in Kosovo had threatened to precipitate a massive humanitarian crisis there.[23] With the establishment of the interim UN administration following the NATO campaign, an uneasy peace prevailed between Serbs and Albanians in the territory for several years until 17 March 2004, when the alleged drowning of three Albanian boys triggered a series of attacks against Serbs and Serb properties across the province. Twenty-one people were killed; more than 700 Serb homes were damaged or destroyed together with thirty-six Serbian Orthodox churches or cultural sites; and more than 4,000 Serbs and other minorities were forced to flee in what appears to have been an orchestrated campaign.[24]

When Kosovo erupted in violence, it was widely known that the frustration within the Kosovo Albanian community was mounting and that continued irresolution of the status question—whether Kosovo would become independent of Serbia or not—could not go on indefinitely. But while deep knowledge of Kosovo alone perhaps could not establish that Kosovo was at the breaking point at that particular time, forensic evidence and other available intelligence

---

[22] Jasmine-Kim Westendorf, *Why Peace Processes Fail: Negotiating Insecurity after Civil War* (Boulder, CO: Lynne Rienner, 2015), 4.

[23] UN Security Council Resolution 1244 (1999) established the interim administration. UN Doc. S/RES/1244 (1999), 10 June 1999.

[24] 'Report of the Secretary-General on the United Nations Interim Administration Mission in Kosovo to the Security Council', UN Doc. S/2004/348, 20 April 2004.

Measuring Peace Consolidation 113

suggested that this was indeed the case. King and Mason mention sightings of paramilitary guerrillas, growing threats of violence, the exploding of an improvised bomb, and a plan which an intelligence agency had discovered for a Kosovo-wide uprising that anticipated the events to come by several months.[25] The difficulty, in part, is that the UN and other peacebuilding organizations often have only limited intelligence capabilities and often are not privy to national intelligence data. Historically there have been concerns, especially on the part of developing countries, about equipping multilateral organizations with intelligence capacities that, it has been feared, could be exploited for illegitimate purposes (notably meddling in domestic affairs). Only gradually are those inhibitions being overcome so that we are seeing the emergence of limited 'peacekeeping intelligence' policy within the UN.[26]

The other challenge that peace operations face with regard to intelligence is the surfeit of information actually or potentially available and the difficulty of establishing the credibility and utility of that information before the fact. 'There is a cacophony of "noise",' Max Hastings has written about intelligence-gathering generally, 'from which "signals"—truths large and small—must be winnowed.'[27] With that objective in mind, the UN and other peacebuilding organizations have been exploring ways of using machine-learning—developing computer programs to use data to learn for themselves—to facilitate the rapid distillation and analysis of large amounts of data in order to analyse trends and patterns in the hope of generating more accurate conflict risk assessments.

PRINCIPLES OF GOOD PRACTICE

An ethnographic approach represents a mode of engagement, a form of orientation, that can underpin good practice. One practice that

---

[25] Iain King and Whit Mason, *Peace at Any Price: How the World Failed Kosovo* (London: C. Hurst & Co., 2006), 6–8.

[26] United Nations, 'DPKO-DFS Policy on Peacekeeping Intelligence', Ref. 2017.07, 2 May 2017.

[27] Max Hastings, 'Smoke and Mirrors, *New York Review of Books*, 27 September 2018, 49.

114 *Measuring Peace*

deserves serious consideration as an integral element of strategic assessment is early and continuous conflict analysis.

Conflict analysis can take many forms.[28] In Chapter 3 we saw the use and value of conflict analysis as developed by the Organization for Security and Co-operation in Europe. The value of conflict analysis can also be appreciated by considering the consequences of its absence. East Timor offers a useful example in this regard. East Timor, we noted, came under UN executive authority from October 1999 until it achieved independence in May 2002. However, even after independence the UN maintained significant peacebuilding engagement in the island-state for several years, including through the brief but quite damaging episode of internal conflict in 2006 that caught the UN unawares. Yet at no point, Olav Ofstad observes in his study of UN peace- and state-building in East Timor, did the UN undertake a conflict analysis and, as a consequence, it was not widely aware of various divisions until after they had manifested themselves in violence. The lesson Ofstad draws from this experience is that 'any peace operation should start with an adequate conflict analysis to ensure understanding of the conflict landscape'.[29] What would such an analysis have entailed in the case of East Timor according to Ofstad?

[a] mapping of all existing conflicts and stakeholders, the threats they represented, the potential for new or changing conflicts and the need for reconciliation and conflict resolution. To provide this, a profound understanding of East Timor's history was required, and of the connections between past and present conflicts. A thorough analysis of relations between different ethnic groups should have been a require-ment, along with analysis of the relationship between political actors and different groups and actors in the informal sector, including the martial arts groups and gangs. The analysis would have needed to include socio-economic, demographic and psychological perspectives. A relevant conflict analysis must also take into account the likely effects of the peace-building process as well as other trends and developments in society.[30]

---

[28] Peter Wallensteen, *Understanding Conflict Resolution*, 4th edn (London: Sage, 2015), ch. 3.

[29] More recently, UN Secretary-General António Gutterres has called for strength-ening capacities to conduct 'conflict or context analysis and to translate analysis into more conflict-sensitive programming'. See 'Peacebuilding and Sustaining Peace', Report of the Secretary-General, UN Doc. A/72/707 and S/2018/43, 18 January 2018, para. 32.

[30] Olav Ofstad, 'Reconciliation and Conflict Resolution in East Timor: Lessons for Future Peace Operations', Oxford Institute for Ethics, Law and Armed Conflict, Working Paper, April 2012, 14.

Measuring Peace Consolidation   115

To this proposed mapping exercise one needs to add environmentally based conflict risks such as actual or potential disputes over natural resources that could cause conflict to re-emerge; natural resources that could be economically significant for conflict activities or that could become the focal point of illegal activities; and environmental contamination that threatens livelihoods.[31]

The point to stress here, however, is not just the importance of conflict analysis but also the importance of continual strategic reassessment throughout the period of engagement. Do the original objectives still support the broad strategic goals of the peacebuilding operation? Have new or unanticipated threats or impediments to a stable peace emerged (e.g., external security challenges, inauspicious political shifts) that require the articulation of new or altered objectives? Has available implementing capacity—internationally and nationally—changed and what implications does this have for meeting an oper-ation's objectives and achieving a sustainable exit? And just as objectives need to be re-evaluated as conditions on the ground change, so will the core tasks associated with these objectives. It is also important to gauge whether the theoretical assumptions underpinning these tasks are sound. For instance, is a weapons buy-back programme actually redu-cing the supply of weapons in a country or merely creating a new regional market for the sale of arms?[32] In sum, a clear road map to a stable peace at the start of an operation is not realistic because circum-stances will evolve, often in unanticipated ways. It is essential, therefore, to reassess the peacebuilding strategy regularly.

There is growing recognition of both the need for continual stra-tegic reassessment and the shortcomings of existing practice. This recognition is reflected in the 2015 Report of the High-Level Inde-pendent Panel on UN Peace Operations:

> Planning must be an ongoing process and informed by objective assess-ments of progress on the ground. The United Nations has not invested

---

[31] As suggested within the context of a UN internal e-discussion: United Nations, 'Consolidated Reply—e-Discussion: 'Measuring Peace Consolidation and Supporting Transition', 9 April and 15 May 2008, http://groups.undp.org/read/messages?id= 245888.

[32] On the ambiguities of disarmament, demobilization, and reintegration 'success', see Claudia Simons and Franzisca Zanker, 'Finding the Cases that Fit: Methodological Challenges in Peace Research', GIGA Working Paper No. 189, German Institute of Global and Area Studies, March 2012, 12.

116   *Measuring Peace*

sufficiently in the monitoring and evaluation of its peace operations or
in building results or impact measurement frameworks for missions to
draw upon. A much stronger results orientation is needed throughout
the planning, implementation and evaluation cycle.[33]

The difficulty in any such analysis, however—and this extends to the
work of all peacebuilding organizations—is to be able to correlate
specific peacebuilding activities to particular effects (results) in the
context of highly complex, open systems in which there will be many
relevant actors as well as factors out of the actors' control.[34]

The experience of benchmarking progress towards peace consolidation, which we discussed in Chapter 3, highlights other elements of
good (and bad) practice that are critical to the effective measurement
of peace consolidation. One is the importance of devising benchmarks or targets that are realistic and measurable. The UN Peacebuilding Commission's 2007 Monitoring and Tracking Mechanism of
the Strategic Framework for Peacebuilding in Burundi, for example,
contained benchmarks that were both unrealistic and too imprecise
to be measured. One such benchmark (of 'good governance') was
'improvement of management of public resources'. Another was the
'existence of a political environment conducive to the peaceful resolution of political conflict through the institutionalization of a culture
and practice of dialogue on major issues and national strategies'—to
be achieved within one year![35]

Precision requires sensitivity to nuance, and this is where the more
differentiated conceptions of peace that we examined in Chapter 1 are
especially salient. As we saw then, scholars have been developing
conceptually rich and rigorous ways of describing the range of post-
conflict conditions, as well as ways of measuring these conditions.
Whether as part of benchmarking or any other assessment exercise,

---

[33] United Nations, 'Uniting Our Strengths for Peace: Politics, Partnerships and
People', Report of the High-Level Independent Panel on United Nations Peace
Operations, UN Doc. A/70/95-S/2015/446, 16 June 2015, para. 172.

[34] For which reason Svein Erik Stave recommends that 'monitoring should
primarily provide on-the-ground information and knowledge of trends that can
inform . . . strategic planning and peacebuilding activities [generally]' and not focus
on the contribution of specific activities. See his 'Measuring Peacebuilding: Challenges, Tools, Actions', NOREF Policy Brief No. 2 (Oslo: Norwegian Peacebuilding
Resource Centre, May 2011), 6.

[35] United Nations Peacebuilding Commission, 'Monitoring and Tracking Mechanism of the Strategic Framework for Peacebuilding in Burundi', UN Doc. PBC/2/
BDI/4, 27 November 2007. I owe this observation to Svein Erik Stave of Fafo.

Measuring Peace Consolidation   117

these conceptualizations can and should be drawn up to provide 'a more fine-grained picture of peace' in societies emerging from war, as Höglund and Söderberg Kovacs put it.[36]

It is also important that any benchmarks and indicators of progress that are adopted are meaningful. They need to focus on concrete outcomes and impacts, not just inputs. If, for instance, it is determined that the establishment and development of a professional, impartial, and independent judiciary is critical for peace consolidation, then one appropriate measure of progress would be whether there is evidence of ethnic or other bias on the part of judges in the performance of their duties, how much trust the court system enjoys among the general public and endangered population groups in particular, and whether judges are respectful of international human rights norms. Measuring how many judges have been trained in human rights law, by contrast, is not a meaningful indicator of the level of competence that the judiciary has attained, although it is a common practice to gravitate towards the quantifiable as a meaningful measure because it is relatively easy 'analysis' to produce.[37]

Incorporating 'local' perspectives into strategic assessment is also important for a number of reasons. One is the simple fact that external peacebuilders tend to be alien to the local culture and are prone to misreading the cues without assistance from the conflict-affected population. Thus to devise meaningful indicators of progress towards a self-sustaining peace, it is important to comprehend what the local anxieties are, what measures are required to address those anxieties, and what evidence of successful address of those anxieties would look like. This is in part what the Everyday Peace Indicators project, discussed in Chapter 3, is seeking to achieve. There is scope for inclusion of so-called 'objective' indicators in such a framework—such as those contained in the Positive Peace Index—but these need to be road-tested for local conditions to ensure that they are relevant.

Inclusion of 'local' perspectives into assessment is also important for generating host country buy-in and participation. Otherwise assessment becomes an alien exercise for those most directly affected. The difficulty that can arise is that there is no singular 'local' perspective.

---

[36] Kristine Höglund and Mimmi Söderberg Kovacs, 'Beyond the Absence of War: The Diversity of Peace in Post-Settlement Societies', *Review of International Studies* 36:2 (2010), 389.

[37] Autesserre discusses this tendency in *Peaceland*, 249–51.

118   *Measuring Peace*

In polarized societies especially, which of course conflict-affected countries often are, there may be several different and mutually exclusive perspectives on the requirements for peace and assessments of progress towards meeting those requirements. One reason for these differences is because interests are often constitutive of perspectives. So when the government of Burundi (one perspective) assesses the progress it has made in promoting security, it may be influenced in its assessment by the natural political drive for self-preservation (i.e., to remain in office or to receive donor assistance for meeting milestones or to discredit the opposition). That is indeed what happened in Burundi where there was a reluctance on the part of the government to recognize the participation of particular youth organizations in politically motivated violence because the government would have had to forswear its association with these organizations if it did.[38] The difficulty arises when differences are so fundamental that local perspectives do not converge at all and it is not possible to achieve a consensus with regard to either assessment or the way forward, as has been the case, for instance, among Croats, Serbs, and Bosniacs in 'post-conflict' Bosnia and Herzegovina.[39]

Stronger strategic assessment is by no means a panacea—for one thing, as discussed below, the knowledge generated can be ignored—but lack of local knowledge has clearly been an impediment to building peace: in Kosovo, where peacebuilders misjudged the 'goodwill' of the Kosovo Liberation Army as partners in democratic statebuilding; in East Timor where UN officials failed to appreciate the various social divisions within the nascent state; and in Afghanistan where the British underestimated the predatory and abusive behaviour of local warlords with whom they chose to ally in reconstruction efforts.

OBSTACLES TO GOOD PRACTICE

There are numerous obstacles to good practice, some more formidable than others. One is the paucity of reliable data. 'The poorer and more disrupted the society, the less likely reliable data will be available

---

[38] As reported to me by a senior official in the UN Peacebuilding Support Office, New York, October 2014.
[39] Christopher Bennett, *Bosnia's Paralysed Peace* (London: C. Hurst & Co., 2016).

Measuring Peace Consolidation 119

for such monitoring,' Ken Menkhaus observes.[40] Most conflict-affected countries are poor countries that, prior to their wars, would have had weak data-collection systems that, since their wars, have been compromised further.[41] There may be few resources to support well-functioning statistical offices and it may still be too dangerous or costly to conduct household surveys. These are countries, moreover, where the writ of the national government may not even extend across the entire country (e.g., Democratic Republic of Congo).[42]

The greatest obstacle to good practice, perhaps, is the tendency to politicize metrics design and reporting. The politicization of metrics design can be inadvertent. For instance, a strategy-based assessment framework will rely on mandates and/or strategic goals by design as the basis for evaluation.[43] While this has one advantage—it means that a peacebuilding organization is being assessed with regard to what it has been tasked to do—the difficulty is that mandates may be adopted for political reasons and therefore may be ambiguous or over/underambitious rather than reflecting the requirements for achieving a consolidated peace.[44]

Politicization of reporting can be a major problem for effective assessment. As one group of NATO analysts observed with regard to ISAF's Strategic Assessment Capability (ISAC) in Afghanistan:

The most prominent obstacle to . . . progress was the tension between ISAC assessments and those produced by HQ ISAF and [Joint Force Commander] Brunssum. The ISAC assessments at this time routinely provided a much less positive outlook than those provided by HQ,

[40] Ken Menkhaus, 'Impact Assessment in Post-Conflict Peacebuilding: Challenges and Future Directions', *Interpeace* (July 2004), 6.

[41] Morten Jerven, *Poor Numbers: How We Are Misled by African Development Statistics and What to Do about It* (Ithaca, NY: Cornell University Press, 2013).

[42] Many organizations are working to develop ways of improving methodologies for measurement and building country statistical capacity, among them the Center for Global Development's Data for African Development programme: https://www.cgdev.org/working-group/data-african-development.

[43] United Nations, *Monitoring Peace Consolidation: United Nations Practitioners' Guide to Benchmarking* (New York: United Nations, 2010), 25–6.

[44] 'Despite the importance of clear and concrete mission objectives, ambiguity does have its role to play in defining the mandate of a peacekeeping mission. It allows [UN Security Council] members to reach consensus (or avoid vetoes) while establishing or renewing peacekeeping missions.' Nathaniel Olin, 'Measuring Peacekeeping: A Review of the Security Council's Benchmarking Process for Peacekeeping Missions', Social Science Research Council, August 2013, 7, https://www.ssrc.org/publications/view/measuring_peacekeeping.

which became increasingly difficult to reconcile. Whilst in theory it was the express aim of an independent assessment to highlight such inconsistencies, it became apparent that this was contentious, especially in the security domain, where instead of being viewed as another aspect of possible "ground truth", the ISAC effort was perceived by some as undermining existing assessment efforts and by extension, the mission itself.

Consequently, ISAC was instructed to discontinue its security domain assessments in 2011.[45]

Sometimes reporting on progress is subordinated to larger political objectives. In the case of the UN Angola Verification Mission, established in May 1991 to verify compliance with the ceasefire arrangements agreed to by the government of Angola and the UNITA rebel group/opposition party, 'the Secretary-General's reports to the Security Council misrepresented the Mission's attribution of certain incidents and otherwise moderated the Mission's account of the extent to which the ceasefire between the Government and UNITA was being breached', according to Rupert Burridge, in his study of the work of the UN operation.[46] Burridge identified at least one instance, on 9 September 1992, when the Secretary-General, in a report to the Security Council, referred to incidents of intimidation and provocation by both government and UNITA supporters when it was clear from the field reports that UNITA was largely to blame for the incidents.[47] The reason for this misrepresentation was that senior UN authorities were concerned not to undermine the political process under way which, it was hoped, would lead to the holding of elections and a successful implementation of the Bicesse peace accords. '[T]he main concern was to report things in a way that the trust and the confidence in the electoral process would not be jeopardized,' the head of Portugal's observer group to the Joint Political-Military Commission, Antonio Monteiro, would later tell Burridge.

---

[45] Nick Lambert, Phil Eles, and Bruce Pennell, 'ISAF Strategic Capability Assessment: A Retrospective', in ISAF Strategic Assessment Capability Workshop, 10–12 December 2014 [2015], version 0.1, para. 3.4.2.

[46] Rupert Burridge, 'A Comparative Analysis of Ceasefire Verification Missions in Angola, Mozambique and Kosovo', unpublished DPhil thesis, Department of Politics and International Relations, University of Oxford.

[47] United Nations Security Council, 'Further Report of the Secretary-General on the United Nations Angola Verification Mission (UNAVEM II)', UN Doc. S/24556, 9 September 1992, para. 6.

Measuring Peace Consolidation   121

As Marrack Goulding, the head of UN peacekeeping operations at the time, would recall in his memoir, 'we calculated that a pessimistic report would increase the risk of disaster'.[48]

It can also be difficult for any organization to evaluate its performance in an entirely disinterested manner. Of course, peacebuilding operations—UN and others—are conducted increasingly in a complex international environment in which other global institutions, regional bodies, the media, non-governmental organizations, and various other agencies monitor and report on these operations. These oversight mechanisms, however, are informal mechanisms that, moreover, vary in their effectiveness. Among other things, these bodies often employ differing criteria in their evaluations, not all of which may be apparent or commensurate with the purposes of strategic assessments. For this reason, it may be advisable for peacebuilding organizations to delegate responsibility for assessment to independent bodies, inside or outside the organization, that would employ relevant criteria. National governments often establish independent, non-partisan agencies—such as the US Government Accountability Office, whose function it is to evaluate how well US governmental policies and programmes are meeting their objectives. Peacebuilding organizations, too, would benefit from independent evaluations of their operations and, in particular, assessments of the extent to which their policies and programmes are working to achieve a consolidated peace.

The greatest risk to sound analysis is wilful disregard. Sometimes an organization may fail to heed the warning signs; other times its member states may fail to take the organization's warning signs on board. In the case of the 1994 Rwandan genocide, both factors were at work. When in December 1993 and January 1994, General Roméo Dallaire, commander of the UN Assistance Mission in Rwanda, warned his superiors in New York that Hutu extremists were planning a campaign to exterminate Tutsis, his requests for more troops and a stronger mandate were denied. As Philip Gourevitch observes, 'no effort was made at peacekeeping headquarters to alert the United Nations Secretariat or the Security Council of the startling news that an "extermination" was reportedly being planned in Rwanda'.[49] After the killing began in April and General Dallaire again pleaded for more

---

[48] Marrack Goulding, *Peacemonger* (London: John Murray, 2002), 187.
[49] Philip Gourevitch, *We Wish to Inform You That Tomorrow We Will Be Killed with Our Families* (New York: Farrar, Strauss, 1998), 105.

122   *Measuring Peace*

support from the UN, his pleas were this time rejected by the UN Security Council, whose members voted instead to reduce the UN peacekeeping force in the country from 2,500 to 270 soldiers.[50] The 100-day genocide that followed would result in the deaths of an estimated 500,000 to 1 million Rwandans, predominantly Tutsis.

*   *   *

Ultimately, the decision to engage, or not, in support of building or maintaining a peace is a political decision on the part of third parties. In an ideal world these decisions would be—and indeed sometimes are—informed by considerations of the requirements for the maintenance of a peace. This chapter has discussed what such considerations would entail—what would help to ensure an effective strategic assessment of peace. It has stressed the value and importance of an 'ethnographic approach' that seeks to gain enhanced understanding of the quality of the peace through knowledge of the local culture, the local history, and especially, the particular conflict dynamics at work in a given conflict. It has also articulated principles of good practice—some of them in evidence in preceding chapters. And, finally, it has discussed some of the obstacles to implementing principles of good practice and how these obstacles might be overcome or at least mitigated.

[50] UN Security Council Resolution 912 (1994), 21 April 1994.

Conclusion

Can we know with any certainty whether a peace is a stable peace? The short answer is 'No'. Conflict dynamics are generally too complex to be able to determine whether in any given case the peace that has been established is a stable peace. We can know with hindsight whether a peace *has been* stable, and we can seek to draw conclusions from historical experience about factors that have contributed to a stable peace—as we did in Chapter 4—but we cannot know with certainty that the peace that prevails currently is a stable peace.

The limitations notwithstanding, it is possible—as the findings of this study suggest—to ascertain the quality of a peace, and the vulnerability of that peace to conflict relapse, with higher levels of confidence. As has been shown in the foregoing chapters—and has been demonstrated with reference to actual practice throughout this book—there are numerous ways for peacebuilding actors to strengthen their capacity to monitor progress towards achieving a consolidated peace. Better assessment, in turn, can inform peacebuilding actors in the reconfiguration and reprioritization of their operations in cases where conditions on the ground have deteriorated (or improved). This is not to suggest that all peace failures can be foreseen or that, if foreseen, can be prevented. The point is simply, but not unimportantly, that more rigorous assessments of the quality of the peace can facilitate more effective external engagement.

There is no single recommended approach to effective strategic assessment. Indeed, one of the more surprising, and most interesting, findings of this research is just how varied strategic assessment can be, whether it is early warning and conflict analysis in the case of the Organization for Security and Co-operation in Europe, benchmarking and co-assessment in the case of the United Nations, or the production of a range of 'peace indicators' in the case of several research institutes, among many other approaches, not all of them represented in these pages. The evidence suggests that many of these initiatives have brought the quality of peace into sharper focus and have helped to improve the efforts of peacebuilding actors.

Varied though the nature of strategic assessment may be, many of the principles that underpin these initiatives are shared principles. What all of them have in common, first and foremost, is a recognition of the vital importance of contextual knowledge for the insights it affords into local conflict-relevant dynamics and their implications for devising appropriate strategies for the maintenance of peace. As a British Academy report on 'Rethinking State Fragility' stressed:

> [I]t is essential to develop a broad and deep understanding of the historic, cultural and political context of a locality, country and region on a case by case basis ... Interventions in the areas of conflict, stability and security are aiming to carry out a transformative political exercise, whatever the mixture of developmental, diplomatic or military purposes. To re-orient the socioeconomic, political and institutional characteristics of a place requires caution, sensitivity, and a depth of knowledge and understanding.[1]

Again, this may seem obvious and, yet, there is a tendency among peacebuilding organizations, to varying degrees, to rely on preconceived models or templates that are often either maladapted or plainly ill-suited to local conditions. The attraction of these templates for practitioners is that they serve to compensate for contextual knowledge that is lacking and, moreover, allow peacebuilding organizations subject to time pressures to respond quickly and deliver results.[2] The lack of contextual knowledge is the chief obstacle to assessing progress towards achieving a consolidated peace.

Other principles of good practice characteristic of sound strategic assessment include early and continuous assessment of conflict and conflict-relevant dynamics. It is important to be sensitive to shifts in

---

[1] Philip Lewis and Helen Wallace, 'Introduction', *Rethinking State Fragility* (London: British Academy, 2015), 2.

[2] Susan L. Woodward, 'Do the Root Causes of Civil War Matter? On Using Knowledge to Improve Peacebuilding Interventions', *Journal of Intervention and Statebuilding* 1:2 (2007), 163; Séverine Autesserre, *Peaceland: Conflict Resolution and the Everyday Politics of International Intervention* (Cambridge: Cambridge University Press, 2014), 13.

Conclusion   125

the conflict landscape—regionally, nationally, and locally—with
regard to key actors, emerging threats, patterns of violence, and
perceptions of security among the population, especially on the part
of any vulnerable communities, whose welfare is often a bellwether
for a stable peace. It is also important to reflect on whether the
theoretical assumptions underpinning peacebuilding tasks are correct
and, if not, to identify what changes are required in the assumptions
and associated tasks. Additionally, it is important that any benchmarks and indicators of progress that are selected to monitor progress
are realistic, measurable, and, above all, meaningful. Finally, it is
critical to incorporate local perspectives into strategic assessments—
in order to achieve local buy-in and to help ensure accuracy. The
difficulty arises when local perspectives diverge—as we should expect
they will in conflict-affected societies. How third parties deal with
contentiousness and contradiction is a topic deserving of greater
attention.

In an ideal world, decisions by third parties regarding engagement
in support of peace would be driven primarily by considerations of
the requirements for the maintenance of peace—in other words,
the state of the peace at stake. Sometimes they are, as we have seen.
But in a world in which states must contend with competing
demands on the use of their scarce resources, and when reputational
costs are at stake, then the decision to engage, the nature of the
engagement, and the duration of the engagement will all be political
decisions ultimately, subject to domestic pressures, budgetary concerns, competing strategic considerations, etc. When China vetoed
the renewal of the United Nations' preventive peacekeeping operation
in Macedonia in 1999, it was not because of careful consideration
of the state of the peace in the former Yugoslav republic, which in fact
would succumb to violent conflict two years later. Rather, it was a
reaction against the establishment of diplomatic relations between
Macedonia and Taiwan.[3]

Better assessments of the quality of peace, therefore, are not a
panacea for conflict recurrence. However, to the extent that sound
analysis can inform policy deliberations, more rigorous assessments
of the robustness of peace have the potential to make a substantial

---

[3] Gabriel Partos, 'What Price Macedonian Peace?' *BBC News*, 25 February 1999,
http://news.bbc.co.uk/1/hi/world/europe/286187.stm.

contribution to the prevention of conflict recurrence. As donor governments, intergovernmental organizations, and non-governmental organizations reassess the nature of their engagement in any given conflict-affected country, they would benefit from greater effort to ascertain the quality of the peace that they have helped to build and its capacity to withstand the pressures to undermine it.

Select Bibliography

Documents

African Union, *Protocol Relating to the Establishment of the Peace and Security Council of the African Union*, Addis Ababa, 2002.

African Union, *Draft Policy Framework for Post-Conflict Reconstruction and Development (PCRD)*, 2006.

Conference on Security and Co-operation in Europe, *Helsinki Final Act*, 1975.

Economic Community of West African States, *The ECOWAS Conflict Prevention Framework*, 2008.

European Commission, *Shared Vision, Common Action: A Stronger Europe*, 2016.

European Commission, 'The former Yugoslav Republic of Macedonia 2016 Report', EC Doc. SWD(2016) 362 final, 9 November 2016.

Government of Sierra Leone, *Interim Poverty Reduction Strategy Paper*, June 2001.

International Development Association, 'Adapting IDA's Performance-Based Allocations to Post-Conflict Countries', May 2001, mimeo.

International Dialogue on Peacebuilding and Statebuilding, 'A New Deal for Engagement in Fragile States', Busan, 30 November 2011.

International Dialogue on Peacebuilding and Statebuilding, 'Progress Report on Fragility Assessments and Indicators', 2012.

International Dialogue on Peacebuilding and Statebuilding, 'Peacebuilding and Statebuilding Indicators: Progress, Interim List and Next Steps', 2013.

International Dialogue on Peacebuilding and Statebuilding, 'New Deal Monitoring Report 2014: Final Version', 2014.

North Atlantic Treaty Organization, *NATO Operations Assessment Handbook*, Interim Version 1.0, 29 January 2011.

Organisation for Economic Co-operation and Development, *Results Based Management in the Development Co-operation Agencies: A Review of Experience* (Paris: OECD, 2000).

Organization for Security and Co-operation in Europe, 'The OSCE Concept of Comprehensive and Co-operative Security', 17 June 2009.

Organization for Security and Co-operation in Europe, 'Ministerial Declaration on the OSCE Corfu Process', Athens, 2 December 2009.

Organization for Security and Co-operation in Europe, 'Background Brief: OSCE Activities and Advantages in the Field of Post-Conflict Rehabilitation', 28 April 2011.

128   *Select Bibliography*

Organization for Security and Co-operation in Europe, Ministerial Council Decision No. 3/11, OSCE Doc. MC.DEC/3/11, 7 December 2011.

Organization for Security and Co-operation in Europe, 'Internal OSCE Open-Ended List of Early Warning Indicators', 23 November 2012.

Organization for Security and Co-operation in Europe, 'Conflict Analysis Toolkit', 2014.

UK Foreign and Commonwealth Office, Ministry of Defence, and Department for International Development, *Building Stability Overseas Strategy*, 2011.

UK Foreign and Commonwealth Office, Ministry of Defence, and Department for International Development, *Building Stability Framework*, 2016.

UK House of Commons, Foreign Affairs Committee, 'Eighth Report, Global Security: Afghanistan and Pakistan', Session 2008–9 (2009).

UN Security Council Resolution 912 (1994), 21 April 1994.

UN Security Council Resolution 1244 (1999), 10 June 1999.

UN Security Council Resolution 1386 (2001), 20 December 2001.

UN Security Council Resolution 1645 (2005), 20 December 2005.

UN Security Council Resolution 1996 (2011), 8 July 2011.

UN Security Council Resolution 2282 (2016), 27 April 2016.

United Nations, 'An Agenda for Peace: Preventive Diplomacy, Peacemaking and Peacekeeping', Report of the Secretary-General pursuant to the statement adopted by the Summit Meeting of the Security Council on 31 January 1992, 17 June 1992.

United Nations, 'Further Report of the Secretary-General on the United Nations Angola Verification Mission (UNAVEM II)', 9 September 1992.

United Nations, 'Final Report of the Secretary-General on the United Nations Observer Mission in Liberia', 12 September 1997.

United Nations, 'Report of the Secretary-General on the United Nations Transitional Administration in East Timor', 26 July 2000.

United Nations, 'No Exit without Strategy: Security Council Decision-Making and the Closure or Transition of United Nations Peacekeeping Operations', Report of the Secretary-General, 20 April 2001.

United Nations, 'Fourteenth Report of the Secretary-General on the United Nations Mission in Sierra Leone', 19 June 2002.

United Nations, 'Fifteenth Report of the Secretary-General on the United Nations Mission in Sierra Leone', 5 September 2002.

United Nations, 'Eighteenth Report of the Secretary-General on the United Nations Mission in Sierra Leone', 23 June 2003.

United Nations, 'Report of the Secretary-General on the United Nations Interim Administration Mission in Kosovo to the Security Council', 20 April 2004.

United Nations, 'Statement by the President of the Security Council', 20 December 2005.

Select Bibliography

United Nations, 'Measuring Peace Consolidation and Supporting Transition', interagency briefing paper prepared for the United Nations Peacebuilding Commission, March 2008.

United Nations, *Monitoring Peace Consolidation: United Nations Practitioners' Guide to Benchmarking*, 2010.

United Nations, 'South Sudan: Planning for Post-CPA UN Presence: Submission to the Policy Committee', undated memorandum (2011).

United Nations, 'Report of the UN Secretary-General on the Situation in the Central African Republic and on the Activities of the United Nations Integrated Peacebuilding Office in That Country', 21 December 2012.

United Nations, 'Report of the Secretary-General on South Sudan', 6 March 2014.

United Nations, 'DPKO-DFS Policy on Peacekeeping Intelligence', Ref. 2017.07, 2 May 2017.

United Nations, 'Monthly Summary of Military and Police Contribution to United Nations Operations', 31 December 2017.

United Nations, 'Peacebuilding and Sustaining Peace', Report of the Secretary-General, 18 January 2018.

United Nations and World Bank, *Pathways for Peace: Inclusive Approaches to Preventing Violent Conflict* (Washington, DC: World Bank, 2018).

United Nations Department of Peacekeeping Operations and Department of Field Support, 'Concept Note: Security Council Benchmarks in the Context of UN Mission Transitions', April 2014.

United Nations General Assembly Resolution 3314 (1974), 14 December 1974.

United Nations General Assembly Resolution 60/180 (2005), 20 December 2005.

United Nations High-Level Independent Panel on United Nations Peace Operations, 'Uniting Our Strengths for Peace: Politics, Partnerships and People', 16 June 2015.

United Nations Office on Drugs and Crime, 'UNODC Global Study on Homicide 2013: Trends, Context, Data', 2014.

United Nations Office of Internal Oversight Services, Inspection and Evaluation Division, 'Inspection Report: OIOS Review of the Relevance, Efficiency and Effectiveness of Results-Based Budgeting of Peacekeeping Operations: Improvements Needed for Realizing Full Potential', Assignment No. IED-08-002, 8 May 2008.

United Nations Peacebuilding Commission, 'Monitoring and Tracking Mechanism of the Strategic Framework for Peacebuilding in Burundi', 27 November 2007.

United Nations Peacebuilding Commission, 'Review of Progress in the Implementation of the Strategic Framework for Peacebuilding in Burundi', 9 July 2008.

130 *Select Bibliography*

United Nations Peacebuilding Community of Practice e-discussion, 'Measuring Peace Consolidation and Supporting Transition: Summary of Responses', 15 May 2008.

United States Agency for International Development, 'The Crisis in Timor-Leste: Causes, Consequences, and Options for Conflict Management and Mitigation', November 2006.

United States Agency for International Development, 'Building Resilience to Recurrent Crisis: USAID Policy and Program Guidance', December 2012.

United States Department of Defense, 'Department of Defense Instruction (Subject: Stability Operations)', Number 3000.05, 16 September 2009.

World Bank, *Articles of Agreement*, 2012.

World Bank, 'Low Income Countries under Stress (LICUS)', FY06-09.

World Bank, *World Development Report 2011: Conflict, Security and Development* (Washington, DC: World Bank, 2011).

World Health Organization, *Basic Documents*, forty-fifth edition, Supplement, October 2006.

Books and Articles

ACCORD, *ACCORD Peacebuilding Handbook*, 2nd edn (Durban: ACCORD, 2015).

Ackermann, Alice, 'Strengthening OSCE Responses to Crises and Conflicts: An Overview', *OSCE Yearbook 2012* (Baden-Baden: Nomos, 2013), 205–11.

Allouche, Jeremy, 'Is It the Right Time for the International Community to Exit Sierra Leone?' Institute of Development Studies Evidence Report No. 38 (November 2013).

'An Editorial', *Journal of Peace Research* 1:1 (1964), 1–4.

Anderson, Royce, 'A Definition of Peace', *Peace and Conflict: Journal of Peace Psychology* 10:2 (2004), 101–16.

Annan, Kofi, 'Closing Remarks to the Conference "Towards a Community of Democracies"', Warsaw, 28 June 2000, UN Press Release SG/SM/7467.

Ansorg, Nadine, Felix Haass, and Julia Strasheim, 'Between Two "Peaces"? Bridging the Gap between Quantitative and Qualitative Conceptualizations in Multi-Method Peace Research', paper presented at the International Studies Association Annual Convention, San Francisco, 3–6 April 2013.

Autesserre, Séverine, 'Construire la paix: conceptions collectives de son établissement, de son maintien et de sa consolidation', *Critique Internationale* 51 (2011), 153–67.

Autesserre, Séverine, *Peaceland: Conflict Resolution and the Everyday Politics of International Intervention* (Cambridge: Cambridge University Press, 2014).

Select Bibliography

Badran, Remzi, 'Intrastate Peace Agreements and the Durability of Peace', *Conflict Management and Peace Science* 31:2 (2014), 193–217.

Banks, Michael, 'Four Conceptions of Peace' in Dennis J. D. Sandole and Ingrid Sandole-Staroste (eds), *Conflict Management and Problem Solving: Interpersonal to International Applications* (London: Frances Pinter, 1987), 259–74.

Barnett, Michael, Hunjoon Kim, Madalene O'Donnell, and Laura Sitea, 'Peacebuilding: What Is in a Name?', *Global Governance* 13:1 (2007), 35–58.

Baumann, Andrea, 'Clash of Organisational Cultures? The Challenge of Integrating Civilian and Military Efforts in Stabilisation Operations', *RUSI Journal* 153:6 (2008), 70–3.

Benner, Thorsten, Stephan Mergenthaler, and Philipp Rotmann, *The New World of UN Peace Operations* (Oxford: Oxford University Press, 2011).

Bennett, Christopher, *Bosnia's Paralysed Peace* (London: C. Hurst & Co., 2016).

Berdal, Mats and Dominik Zaum, 'Power after Peace' in Mats Berdal and Dominik Zaum (eds), *Political Economy of Statebuilding: Power after Peace* (Abingdon: Routledge, 2013).

Bhuta, Nehal, 'Governmentalizing Sovereignty: Indexes of State Fragility and the Calculability of Political Order' in Kevin E. Davis, Angelina Fisher, Benedict Kingsbury, and Sally Engle Merry (eds), *Governance by Indicators: Global Power through Quantification and Rankings* (Oxford: Oxford University Press, 2012).

Bildt, Carl, *Peace Journey: The Struggle for Peace in Bosnia* (London: Weidenfield and Nicholson, 1998).

Blechman, Barry M., William J. Durch, Wendy Eaton, and Julie Werbel, *Effective Transitions from Peace Operations to Sustainable Peace: Final Report* (Washington, DC: DFI International, September 1997).

Boulding, Kenneth, *Stable Peace* (Austin, TX: University of Texas Press, 1978).

Box-Steffensmeier, Janet M. and Bradford S. Jones, *Event History Modeling: A Guide for Social Scientists* (New York: Cambridge University Press, 2004).

Boyle, Michael J., *Violence after War: Explaining Instability in Post-Conflict States* (Baltimore: Johns Hopkins University Press, 2014).

Bratt, Duane, 'Assessing the Success of UN Peacekeeping Operations', *International Peacekeeping* 3:4 (1996), 64–81.

Brauman, Rony, 'Médecins sans Frontières and the ICRC: Matters of Principle', *International Review of the Red Cross* 94:888 (Winter 2012), 1523–35.

British Academy, *Rethinking State Fragility* (London: British Academy, 2015).

Brown, Michael E. (ed.), *The International Dimensions of Internal Conflict* (Cambridge, MA: MIT Press, 1996).

132   *Select Bibliography*

Bufacci, Vittorio, 'Two Concepts of Violence', *Political Studies Review* 3 (2005), 193–204.

Burridge, Rupert, 'A Comparative Analysis of Ceasefire Verification Missions in Angola, Mozambique and Kosovo', unpublished DPhil thesis, Department of Politics and International Relations, University of Oxford.

Call, Charles T., 'Knowing Peace When You See It: Setting Standards for Peacebuilding Success', *Civil Wars* 10:2 (2008), 173–94.

Call, Charles T., *Why Peace Fails: The Causes and Prevention of Civil War Recurrence* (Washington, DC: Georgetown University Press, 2012).

Call, Charles T. and Cedric de Coning, 'Conclusion: Are Rising Powers Breaking the Peacebuilding Mold?' in Charles T. Call and Cedric de Coning (eds), *Rising Powers and Peacebuilding: Breaking the Mold?* (Basingstoke: Palgrave Macmillan: 2017).

Call, Charles T. and Elizabeth M. Cousens, 'Ending Wars and Building Peace: International Responses to War-Torn Societies', *International Studies Perspectives* 9:1 (2008), 1–21.

Call, Charles T. with Vanessa Wyeth (eds), *Building States to Build Peace* (Boulder, CO: Lynne Rienner, 2008).

Cammett, Melani and Edmund Malesky, 'Power Sharing in Postconflict Societies: Implications for Peace and Governance', *Journal of Conflict Resolution* 56:6 (2012), 982–1016.

Campbell, Susanna P. with Leonard Kayobera and Justine Nkurunziza, 'Independent External Evaluation: Peacebuilding Projects in Burundi', March 2010.

Caplan, Richard, *International Governance of War-Torn Territories: Rule and Reconstruction* (Oxford: Oxford University Press 2005).

Caplan, Richard (ed.), *Exit Strategies and State Building* (New York: Oxford University Press, 2012).

Caplan, Richard and Anke Hoeffler, 'Why Peace Endures: An Analysis of Post-Conflict Peace Stabilization', *European Journal of International Security* 2:2 (2017), 133–52.

Cederman, Lars-Erik and Manuel Vogt, 'Dynamics and Logics of Civil War', *Journal of Conflict Resolution* 61:9 (2017), 1992–2016.

Center for Systemic Peace, *State Fragility Index and Matrix 2016*, https://www.systemicpeace.org/inscr/SFImatrix2016c.pdf.

Chandler, David, 'Resilience and Human Security: The Post-interventionist Paradigm', *Security Dialogue* 43:3 (2012), 213–29.

Cleves, Mario, Roberto G. Gutierrez, William Gould, and Yulia V. Marchenko, *An Introduction to Survival Analysis Using Stata*, 3rd edn (College Station, TX: Stata Press, 2010).

Coady, C. A. J., 'The Idea of Violence', *Journal of Applied Philosophy* 3:1 (1986), 3–19.

Select Bibliography   133

Collier, Paul and Anke Hoeffler, 'Greed and Grievance in Civil War', *Oxford Economic Papers* 56:4 (2004), 563–95.

Collier, Paul, Anke Hoeffler, and Måns Söderbom, 'Post-Conflict Risks', *Journal of Peace Research* 45:4 (2008), 461–78.

Collier, Paul, Anke Hoeffler, and Dominic Rohner, 'Beyond Greed and Grievance: Feasibility and Civil War', *Oxford Economic Papers* 61 (2009), 1–27.

Connable, Ben, *Embracing the Fog of War: Assessments and Metrics in Counterinsurgency* (Santa Monica, CA: RAND Corporation, 2012).

Cooley, Alexander and Jack Snyder (eds), *Ranking the World: Grading States as a Tool of Global Governance* (Cambridge: Cambridge University Press, 2015).

Davis, Kevin E., Benedict Kingsbury, and Sally Engle Merry, 'Introduction: Global Governance by Indicators' in Kevin E. Davis, Angelina Fisher, Benedict Kingsbury, and Sally Engle Merry (eds), *Governance by Indicators: Global Power through Quantification and Rankings* (Oxford: Oxford University Press, 2012).

De Vries, W. F. M., 'Meaningful Measures: Indicators on Progress, Progress on Indicators', *International Statistical Review* 69:2 (2001), 313–31.

Del Castillo, Graciana, *Rebuilding War-Torn States: The Challenge of Post-Conflict Economic Reconstruction* (New York: Oxford University Press, 2008).

Diehl, Paul F. and Daniel Druckman, 'Evaluating Peace Operations' in Joachim A. Koops, Noorie MacQueen, Thierry Tardy, and Paul D. Williams (eds), *The Oxford Handbook of United Nations Peacekeeping Operations* (Oxford: Oxford University Press, 2015).

Doyle, Michael W. and Nicholas Sambanis, 'International Peacebuilding: A Theoretical and Quantitative Analysis', *American Political Science Review* 94:4 (2000), 779–801.

Doyle, Michael W. and Nicholas Sambanis, *Making War and Building Peace: United Nations Peace Operations* (Princeton, NJ: Princeton University Press, 2006).

Doyle, Michael W., Ian Johnstone, and Robert C. Orr, *Keeping the Peace: Multidimensional UN Operations in Cambodia and El Salvador* (Cambridge: Cambridge University Press, 1997).

Economic Strategy and Project Identification Group, 'Towards a Kosovo Development Plan: The State of the Kosovo Economy and Possible Ways Forward', ESPIG Policy Paper No. 1, Pristina, 24 August 2004.

Eikenberry, Karl and Stephen D. Krasner (eds), special issue on 'Civil Wars and Global Disorder: Threats and Opportunities', *Dædalus* 146: (2017).

Farrell, Theo, *Unwinnable: Britain's War in Afghanistan 2001–2014* (London: Bodley Head, 2017).

134 *Select Bibliography*

Fast, Larissa, 'Culture Clash: A Humanitarian Perspective on Civil-Military Relations', *Peace Policy* (2010), http://peacepolicy.nd.edu/2010/04/09/culture-clash-a-humanitarian-perspective-on-civil-military-interactions.

Fearon, James D. and David D. Laitin, 'Ethnicity, Insurgency, and Civil War', *American Political Science Review* 97:1 (2003), 75–90.

Firchow, Pamina and Roger Mac Ginty, 'Measuring Peace: Comparability, Commensurability and Complementarity Using Bottom-Up Indicators', *International Studies Review* 19:1 (2017), 6–27.

Florea, Adrian, 'Where Do We Go from Here? Conceptual, Theoretical, and Methodological Gaps in the Large-N Civil War Research Program', *International Studies Review* 14:1 (2012), 78–98.

Fortna, Virginia Page, 'Does Peacekeeping Keep Peace? International Intervention and the Duration of Peace after Civil War', *International Studies Quarterly* 48:2 (2004), 269–92.

Fortna, Virginia Page, *Does Peacekeeping Work? Shaping Belligerents' Choices after Civil War* (Princeton, NJ: Princeton University Press, 2008).

Franke, Benedikt, *Security Cooperation in Africa: A Reappraisal* (Boulder, CO: Lynne Rienner, 2009).

Fund for Peace, *Failed States Index 2005*, http://foreignpolicy.com/2009/10/22/the-failed-states-index-2005.

Fund for Peace, *CAST: Conflict Assessment Framework Manual* (2014), http://library.fundforpeace.org/library/cfsir1418-castmanual2014-english-03a.pdf.

Fund for Peace, *Fragile States Index 2016*, http://fsi.fundforpeace.org.

Fund for Peace, *Fragile States Index 2017*, http://fundforpeace.org/fsi.

Gagnon, Chip and Keith Brown (eds), *Post-Conflict Studies: An Interdisciplinary Approach* (Abingdon: Routledge, 2014).

Gallie, W. B., 'Essentially Contested Concepts', *Proceedings of the Aristotelian Society*, New Series 56 (1956), 167–98.

Galtung, Johan, 'Violence, Peace, and Peace Research', *Journal of Peace Research* 6:3 (1974), 167–91.

Galtung, Johan, 'Twenty-Five Years of Peace Research: Ten Challenges and Some Responses', *Journal of Peace Research* 22:2 (1985), 141–58.

Galtung, Johan, 'Cultural Violence', *Journal of Peace Research* 27:3 (1990), 291–305.

Gates, Scott, Håvard Hegre, Mark Jones, and Håvard Strand, 'Institutional Inconsistency and Political Instability: Polity Duration, 1800–2000', *American Journal of Political Science* 50:4 (2006), 893–908.

Ghani, Ashraf and Clare Lockhart, *Fixing Failed States* (New York: Oxford University Press, 2009).

Gilligan, Michael J. and Ernest J. Sergenti, 'Do UN Interventions Cause Peace? Using Matching to Improve Causal Inference', *Quarterly Journal of Political Science* 3:2 (2008), 89–122.

Select Bibliography

Gleditsch, Nils Petter, Peter Wallensteen, Mikael Eriksson, Margareta Sollenberg, and Hävard Strand, 'Armed Conflict, 1946–2001: A New Dataset', *Journal of Peace Research* 39:5 (2002), 615–37.

Goldstone, Jack, Jonathan Haughton, Karol Soltan, and Clifford Zunes, 'Strategy Framework for the Assessment and Treatment of Fragile States', USAID PPC/IDEAS and Center for Institutional Reform and the Informal Sector (IRIS), Washington, DC, November 2003.

Goldstone, Jack, Robert H. Bates, David L. Epstein, Ted Robert Gurr, Michael B. Lustik, Monty G. Marshall, Jay Ulfelder, and Mark Woodward, 'A Global Model for Forecasting Political Instability', *American Journal of Political Science* 54:1 (2010), 190–208.

Goulding, Marrack, *Peacemonger* (London: John Murray, 2002).

Gourevitch, Philip, *We Wish to Inform You That Tomorrow We Will Be Killed with Our Families* (New York: Farrar, Strauss, 1998).

Gowan, Richard, 'Happy Birthday UN: The Peacekeeping Quagmire', *Georgetown Journal of International Affairs* 16:2 (12 August 2015).

Guarrieri, Thomas R., A. Cooper Drury, and Amanda Murdie, 'Introduction: Exploring Peace', *International Studies Review* 19 (2017), 1–5.

Guelke, Adrian, 'Brief Reflections on Measuring Peace', *Shared Space: A Research Journal on Peace, Conflict and Community Relations in Northern Ireland* 18 (2014), 105–12.

Gurr, Ted Robert, *Why Men Rebel* (Princeton, NJ: Princeton University Press, 1970).

Hartzell, Caroline and Matthew Hoddie, 'Institutionalizing Peace: Power Sharing and Post-Civil War Conflict Management', *American Journal of Political Science* 47:2 (2003), 318–32.

Hartzell, Caroline, Matthew Hoddie, and Donald Rothchild, 'Stabilizing the Peace after Civil War: An Investigation of Some Key Variables', *International Organization* 55:1 (2001), 183–208.

Hastings, Max, 'Smoke and Mirrors, *New York Review of Books*, 27 September 2018, 49–51.

Hegre, Håvard, Tanja Ellingsen, Scott Gates, and Nils Petter Gleditsch, 'Toward a Democratic Civil Peace? Democracy, Political Change, and Civil War 1816–1992', *American Political Science Review* 95:1 (2001), 17–33.

Hegre, Håvard, Lisa Hultman, and Håvard Mokleiv Nygård, 'Evaluating the Conflict-Reducing Effect of UN Peacekeeping Operations', mimeo (2014), https://www.dropbox.com/s/m1k612fg8vg1syc/PKO_prediction_2013.pdf.

Hegre, Håvard, Lisa Hultman, and Håvard Mokleiv Nygård, 'Peacekeeping Works: An Assessment of the Effectiveness of UN Peacekeeping', *Conflict Trends* 01/2015, Peace Research Institute Oslo, 2015.

Hoeffler, Anke, 'Can International Interventions Secure the Peace?' *International Area Studies Review* 17:1 (2014), 75–94.

136   *Select Bibliography*

Höglund, Kristine and Mimmi Söderberg Kovacs, 'Beyond the Absence of War: The Diversity of Peace in Post-Settlement Societies', *Review of International Studies* 36:2 (2010), 367–90.

Hollis, Martin and Steven Smith, *Explaining and Understanding International Relations* (Oxford: Clarendon Press, 1991).

Howard, Lise Morjé, *UN Peacekeeping in Civil Wars* (New York: Cambridge University Press, 2008).

Howard, Lise Morjé, *Power in Peacekeeping* (New York: Cambridge University Press, 2019).

Hultman, Lisa, Jacob D. Kathman, and Megan Shannon, 'United Nations Peacekeeping Dynamics and the Duration of Post-Civil Conflict Peace', *Conflict Management and Peace Science* 33:3 (2016), 231–49.

Institute for Economics and Peace, *Positive Peace Report 2016* (Sydney: IEP, 2016).

International Alert, *Programming Framework for International Alert: Design, Monitoring and Evaluation* (London: International Alert, 2010).

International Peace Institute, *IPI Peacekeeping Database*, www.providingforpeacekeeping.org database.

International Peace Institute, 'Catalogue of Indices 2016: Data for a Changing World', 28 September 2016, https://theglobalobservatory.org/2016/09/catalogue-indices.

Ishiyama, John and Anna Batta, 'Rebel Organizations and Conflict Management in Post-Conflict Societies 1990–2009', *Civil Wars* 13:4 (2011), 437–57.

Ishizuka, Katsumi, *The History of Peace-Building in East Timor: The Issues of International Intervention* (Cambridge: Cambridge University Press, 2010).

Jenkins, Rob, *Peacebuilding: From Concept to Commission* (Abingdon: Routledge, 2013).

Jerven, Morten, *Poor Numbers: How We Are Misled by African Development Statistics and What to Do about It* (Ithaca, NY: Cornell University Press, 2013).

Jones, Will, Ricardo Soares de Oliveira, and Harry Verhoeven, 'Africa's Illiberal State-Builders', Working Paper 89, Refugee Studies Centre, University of Oxford, 2013.

Keane, John, *Reflections on Violence* (London: Verso, 1996).

Keen, David, 'War and Peace: What's the Difference', *International Peacekeeping* 7:4 (2000), 1–22.

Kelley, Judith G. and Beth A. Simmons, 'Politics by Number: Indicators as Social Pressure in International Relations', *American Journal of Political Science* 59:1 (2015), 55–70.

Kemp, Walter, 'The OSCE and the Management of Ethnopolitical Conflict' in Stefan Wolff and Marc Weller (eds), *Institutions for the Management of*

Select Bibliography

Ethnopolitical Conflicts in Eastern and Central Europe (Strasbourg: Council of Europe Publishing, 2008).

Kier, Elizabeth, *Imagining War: French and British Military Doctrine between the Wars* (Princeton, NJ: Princeton University Press, 1997).

Kilcullen, David, *The Accidental Guerrilla: Fighting Small Wars in the Midst of a Big One* (London: C. Hurst & Co., 2017).

King, Charles, 'The Decline of International Studies', *Foreign Affairs* 94:4 (July–August 2015), 88–98.

King, Iain and Whit Mason, *Peace at Any Price: How the World Failed Kosovo* (London: C. Hurst & Co., 2006).

Klein, James P., Gary Goertz, and Paul F. Diehl, 'The Peace Scale: Conceptualizing and Operationalizing Non-Rivalry and Peace', *Conflict Management and Peace Science* 25 (2008), 67–80.

Kreutz, Joakim, 'How and When Armed Conflicts End: Introducing the UCDP Conflict Termination Dataset', *Journal of Peace Research* 47:2 (2010), 243–50.

Kreutz, Joakim, UCDP Conflict Termination Dataset Codebook, v.2–2015, 19 February 2016.

Kurtz, Jon, 'Resilience', United States Institute of Peace *Insights* (Summer 2014).

Lacan, Jacques, *Écrits* (Paris: Éditions du Seuil, 1966).

Lamb, Robert D., Kathryn Mixon, and Sarah Minot, *The Uncertain Transition from Stability to Peace* (Washington, DC: Center for Strategic and International Studies, February 2015).

Lambert, Nicholas J., 'Measuring the Success of the NATO Operation in Bosnia and Herzegovina 1995–2000', *European Journal of Operational Research* 140 (2002), 459–81.

Lambert, Nick, Phil Eles, and Bruce Pennell, 'ISAF Strategic Assessment Capability: A Retrospective', in ISAF Strategic Assessment Capability: Final Workshop, 10–12 December 2014 (2015), 14–25.

Landgren, Karin, 'Unmeasured Positive Legacies of UN Peace Operations', *International Peacekeeping* (forthcoming).

Lansford, Jennifer E. and Kenneth A. Dodge, 'Cultural Norms for Adult Corporal Punishment of Children and Societal Rates of Endorsement and Use of Violence', *Parenting, Science and Practice* 8:3 (2008), 257–70.

Lederach, John Paul, *Building Peace: Sustainable Reconciliation in Divided Societies* (Washington, DC: United States Institute of Peace Press, 1967).

Licklider, Roy A. (ed.), *Stopping the Killing: How Civil Wars End* (New York: New York University Press, 1993).

Licklider, Roy A., 'The Consequences of Negotiated Settlement in Civil Wars 1945–1993', *American Political Science Review* 89:3 (1995), 681–90.

Linz, Juan J., *Totalitarian and Authoritarian Regimes* (Boulder, CO: Lynne Rienner, 2000).

Select Bibliography

Linz, Juan J. and Alfred Stepan, *Problems of Democratic Transition and Consolidation: Southern Europe, South America and Post-Communist Europe* (Baltimore: Johns Hopkins University Press, 1996).

Lipson, Michael, 'Performance under Ambiguity: International Organization Performance in UN Peacekeeping', *Review of International Organizations* 5:3 (2010), 249–84.

Lund, Michael, 'What Kind of Peace Is Being Built? Taking Stock of Post-Conflict Peacebuilding and Charting Future Directions', mimeo, January 2003.

Mac Ginty, Roger, 'Indicators +: A Proposal for Everyday Peace Indicators', *Evaluation and Program Planning* 36 (2013), 56–63.

MacFarlane, S. Neil and Yuen Foong Khong, *Human Security and the United Nations: A Critical History* (Bloomington, IN: Indiana University Press, 2006).

Mansfield, Edward D. and Jack Snyder, *Electing to Fight: Why Emerging Democracies Go to War* (Cambridge, MA: MIT Press, 2007).

March, James G. and Johan P. Olsen, 'The Logic of Appropriateness' in Robert E. Goodin, Michael Moran, and Martin Rein (eds), *The Oxford Handbook of Public Policy* (Oxford: Oxford University Press, 2008), 689–708.

Marshall, Monty G. and Gabrielle Elzinga-Marshall, 'Global Report 2017: Conflict, Governance, and Fragility', http://www.systemicpeace.org/globalreport.html.

Martin, Philip, 'Coming Together: Power-Sharing and the Durability of Negotiated Peace Settlements', *Civil Wars* 15:3 (2013), 332–58.

Mason, David T., Mehmet Gurses, Patrick T Brandt, and Jason Michael Quinn, 'When Civil Wars Recur: Conditions for Durable Peace after Civil Wars', *International Studies Perspectives* 12:2 (2011), 171–89.

McCandless, Erin, 'Wicked Problems in Peacebuilding and Statebuilding: Making Progress in Measuring Progress through the New Deal', *Global Governance* 19 (2013), 227–48.

McNamara, Robert, *In Retrospect: The Tragedy and Lessons of Vietnam* (New York: Times Books, 1995).

Menkhaus, Ken, 'Impact Assessment in Post-Conflict Peacebuilding: Challenges and Future Directions', *Interpeace* (July 2004).

Menkhaus, Ken, 'Making Sense of Resilience in Peacebuilding Contexts: Approaches, Applications, Implications', *Geneva Peacebuilding Platform Paper* 6 (2013).

Millar, Gearoid, Jair van der Lijn, and Willemijn Verkoren, 'Peacebuilding Plans and Local Reconfigurations: Frictions between Imported Processes and Indigenous Practices', *International Peacekeeping* 20:2 (2013), 137–43.

Mitchell, Chris, 'Recognising Conflict' in Tom Woodhouse (ed.), *Peacemaking in a Troubled World* (New York/Oxford: Berg, 1991), 209–25.

Select Bibliography

Neukirch, Claus, 'Early Warning and Early Action: Current Developments in OSCE Conflict Prevention Activities' in *OSCE Yearbook 2013* (Baden-Baden: Nomos, 2014), 123–33.

Nilsson, Desirée, 'Partial Peace: Rebel Groups Inside and Outside of Civil War Settlements', *Journal of Peace Research* 45:4 (2008), 479–95.

Nilsson, Desirée, 'Anchoring the Peace: Civil Society Actors in Peace Accords and Durable Peace', *International Interactions: Empirical and Theoretical Research in International Relations* 38:2 (2012), 243–66.

Ofstad, Olav, 'Reconciliation and Conflict Resolution in East Timor: Lessons for Future Peace Operations', Oxford Institute for Ethics, Law and Armed Conflict, Working Paper, April 2012.

Olin, Nathaniel, 'Measuring Peacekeeping: A Review of the Security Council's Benchmarking Process for Peacekeeping Missions,' Social Science Research Council, August 2013, https://www.ssrc.org/publications/view/measuring_peacekeeping.

Olonisakin, 'Funmi', *Peacekeeping in Sierra Leone: The Story of UNAMSIL* (New York: International Peace Academy, 2008).

Owen, William J. and Stephan Flemming, 'Perspectives on the NATO Success Measurement Systems: The Record and the Way Forward', Workshop Proceedings from the Cornwallis Group VII: Analysis for Compliance and Peace Building, Ottawa, Canada, 25–28 March 2002, http://www.ismor.com/cornwallis/workshop_2002.shtml.

Paris, Roland, 'Peacekeeping and the Constraints of Global Culture', *European Journal of International Relations* 9:3 (2003), 441–73.

Paris, Roland, *At War's End: Building Peace after Civil Conflict* (Cambridge: Cambridge University Press, 2004).

Partos, Gabriel, 'What Price Macedonian Peace?' *BBC News*, 25 February 1999.

Pettersson, Therése and Peter Wallensteen, 'Armed Conflicts, 1946–2014', *Journal of Peace Research* 52:4 (2015), 536–50.

Posen, Barry R., 'The Security Dilemma and Ethnic Conflict', *Survival* 35:1 (1993), 27–47.

Price, Megan, *Measuring Security Progress: Politics, Challenges and Solutions* (The Hague: Clingendael Institute, 2015).

Pugh, Michael, 'The Political Economy of Exit' in Richard Caplan (ed.), *Exit Strategies and State Building* (New York: Oxford University Press, 2012).

Putzel, James and Jonathan Di John, 'Meeting the Challenges of Crisis States', Crisis States Research Centre Report, London School of Economics and Political Science, 2012.

Race, Jeffrey, *War Comes to Long An: Revolutionary Conflict in a Vietnamese Province*, 2nd edn (Berkeley, CA: University of California Press, 2010).

Rees, Edward, 'Under Pressure. FALINTIL–Forças De Defesa De Timor-Leste: Three Decades of Defence Force Development in Timor-Leste

140 *Select Bibliography*

1975–2004'. Working Paper, no. 139. Geneva Centre for the Democratic Control of Armed Forces, April 2004.

Richards, Paul, 'New War: An Ethnographic Approach' in Paul Richards (ed.), *No Peace, No War: An Anthropology of Contemporary Armed Conflicts* (Athens, OH/Oxford: Ohio University Press/James Currey, 2005).

Richmond, Oliver P., *The Transformation of Peace* (Basingstoke: Palgrave Macmillan, 2005).

Richmond, Oliver P., 'Becoming Liberal, Unbecoming Liberalism: Liberal–Local Hybridity via the Everyday as a Response to the Paradoxes of Liberal Peacebuilding', *Journal of Intervention and Statebuilding* 3:3 (2009), 324–44.

Richmond, Oliver P., *Failed Statebuilding: Intervention and the Dynamics of Peace Formation* (New Haven, CT: Yale University Press, 2014).

Rotmann, Philipp and Léa Steinacker, *Stabilization: Doctrine, Organisation and Practice* (Berlin: Global Public Policy Institute, December 2013).

Rudloff, Peter and Michael G. Findley, 'The Downstream Effects of Combatant Fragmentation on Civil War Recurrence', *Journal of Peace Research* 53:1 (2016), 19–32.

Sambanis, Nicholas, 'What Is Civil War? Conceptual and Empirical Complexities of an Operational Definition', *Journal of Conflict Resolution* 48:6 (2004), 814–58.

Sarkees, Meredith Reid, Codebook for the Intra-State Wars v.4.0: Definitions and Variables, 1–2, http://www.correlatesofwar.org/data-sets/COW-war.

Sarkees, Meredith Reid and Frank Whelon Wayman, *Resort to War: A Data Guide to Inter-State, Extra-State, Intra-state, and Non-State Wars, 1816–2007* (Washington, DC: CQ Press, 2010).

Scambary, James, 'Anatomy of a Conflict: The 2006–2007 Communal Violence in East Timor', *Conflict, Security and Development* 9:2 (2009), 265–88.

Schroden, Jonathan, 'Operations Assessment at ISAF: Changing Paradigms' in Andrew Williams, James Bexfield, Fabrizio Fitzgerald Farina, and Johannes de Nijs (eds), *Innovations in Operations Assessment: Recent Developments in Measuring Results in Conflict Environments* (The Hague: NATO Communications and Information Agency, 2014).

Simons, Claudia and Franzisca Zanker, 'Finding the Cases that Fit: Methodological Challenges in Peace Research', GIGA Working Paper No. 189, German Institute of Global and Area Studies, March 2012.

Sriram, Chandra Lekha, 'Justice as Peace? Liberal Peacebuilding and Strategies of Transitional Justice', *Global Society* 21:4 (2007), 579–91.

Stave, Svein Erik, 'Measuring Peacebuilding: Challenges, Tools, Actions', NOREF Policy Brief No. 2 (Oslo: Norwegian Peacebuilding Resource Centre, May 2011).

Select Bibliography

Stewart, Frances (ed.), *Horizontal Inequalities and Conflict: Understanding Group Violence in Multiethnic Societies* (Basingstoke: Palgrave Macmillan, 2008).

Stewart, Rory and Gerald Knaus, *Can Intervention Work?* (New York: W. W. Norton & Co., 2011).

Stiglitz, Joseph, 'Towards a Better Measure of Well-Being', *Financial Times*, 14 September 2009.

Stiglitz, Joseph E., Amartya Sen, and Jean-Paul Fitoussi, *Report by the Commission on the Measurement of Economic Performance and Social Progress* (2009), http://www.stiglitz-sen-fitoussi.fr.

Street, Anne M., Howard Mollett, and Jennifer Smith, 'Experiences of the United Nations Peacebuilding Commission in Sierra Leone and Burundi', *Journal of Peacebuilding and Development* 4:2 (2008), 33–46.

Suhrke, Astri, *When More Is Less: The International Project in Afghanistan* (London: C. Hurst & Co., 2011).

Suhrke, Astri and Mats Berdal (eds), *The Peace in Between: Post-War Violence and Peacebuilding* (Abingdon: Routledge, 2012).

Suhrke, Astri and Ingrid Samset, 'What's in a Figure? Estimating Recurrence of Civil War', *International Peacekeeping* 14:2 (2007), 195–203.

Themnér, Lotta and Peter Wallensteen, 'Armed Conflicts: 1946–2011', *Journal of Peace Research* 49:4 (2012), 565–75.

Themnér, Lotta and Peter Wallensteen, 'Armed Conflicts, 1946–2013', *Journal of Peace Research* 51:4 (2014), 541–54.

Toft, Monica Duffy, *Securing the Peace: The Durable Settlement of Civil Wars* (Princeton, NJ: Princeton University Press, 2009).

Toft, Monica Duffy, 'Ending Civil Wars: A Case for Rebel Victory?' *International Security*, 34:4 (2010), 7–36.

Uppsala Conflict Data Program and Peace Research Institute, Oslo, 'UCDP/ PRIO Armed Conflict Dataset v.4-2014a, 1946–2013'.

Uppsala Conflict Data Program and Peace Research Institute, Oslo, 'UCDP/ PRIO Armed Conflict Dataset Codebook', Version 4-2014a.

Von der Schulenburg, Michael, *Rethinking Peacebuilding: Transforming the UN Approach* (New York: International Peace Institute, September 2014).

Wæver, Ole, 'Peace and Security: Two Concepts and Their Relationship' in Stefano Guzzini and Dietrich Jung (eds), *Contemporary Security Analysis and Copenhagen Peace Research* (London: Routledge, 2004), 53–65.

Wallensteen, Peter, *Quality Peace: Peacebuilding, Victory and World Order* (New York: Oxford University Press, 2015).

Wallensteen, Peter, *Understanding Conflict Resolution*, 4th edn (London: Sage, 2015).

Walter, Barbara F., 'Conflict Relapse and the Sustainability of Post-Conflict Peace', background paper to World Bank, *World Development Report 2011*, 13 September 2010, openknowledge.worldbank.org/handle/10986/9069.

142 *Select Bibliography*

Walter, Barbara F., 'Why Bad Governance Leads to Repeat Civil War', *Journal of Conflict Resolution* 59:7 (2015), 1242–72.

Walter, Barbara F. and Jack L. Snyder (eds), *Civil Wars, Insecurity, and Intervention* (New York: Columbia University Press, 1999).

Waltz, Kenneth N. 'The Stability of a Bipolar World', *Dædalus* 93:3 (1964), 881–909.

Weber, Max, 'Social Psychology of the World's Religions' in *From Max Weber: Essays in Sociology*, eds. H. H. Gerth and C. Wright Mills (New York: Oxford University Press, 1958 [1913]).

Weidmann, Nils, Jan Ketil Rød, and Lars-Erik Cederman, 'Representing Ethnic Groups in Space: A New Dataset', *Journal of Peace Research* 47:4 (2010), 491–9.

Westendorf, Jasmine-Kim, *Why Peace Processes Fail: Negotiating Insecurity after Civil War* (Boulder, CO: Lynne Rienner, 2015).

Westerwinter, Oliver, 'The Informal Powers of International Secretariats: Informal Governance and Brokerage in United Nations Post-Conflict Peacebuilding', paper presented at Nuffield College, Oxford, 6 December 2013.

Williams, Andrew, James Bexfield, Fabrizio Fitzgerald Farina, and Johannes de Nijs (eds), *Innovation in Operations Assessment: Recent Developments in Measuring Results in Conflict Environments* (The Hague: NATO Information and Communications Agency, 2014).

Williams, Paul D, 'The African Union's Conflict Management Capabilities', Council on Foreign Relations Working Paper (New York: Council on Foreign Relations, October 2011).

Wilton Park, 'Measuring Peace Consolidation', report of conference held at Wilton House, Wilton Park, UK, 15–17 October 2014.

Woods, Ngaire, *The Globalizers: The IMF, the World Bank and Their Borrowers* (Ithaca, NY: Cornell University Press, 2006).

Woodward, Susan L., 'Do the Root Causes of Civil War Matter? On Using Knowledge to Improve Peacebuilding Interventions', *Journal of Intervention and Statebuilding* 1:2 (2007), 143–70.

Woodward, Susan L., 'The IFIs and Post-Conflict Political Economy' in Mats Berdal and Dominik Zaum (eds), *Political Economy of Statebuilding: Power after Peace* (Abingdon: Routledge, 2013).

Wucherpfennig, Julian, Nils W. Metternich, Lars-Erik Cederman, and Kristian Skrede Gleditsch, 'Ethnicity, the State, and the Duration of Civil War', *World Politics* 64:1 (2012), 79–115.

Wyeth, Vanessa, 'Knights in Fragile Armour: The Rise of the g7+', *Global Governance* 18:1 (2012), 7–12.

Yamaguchi, Masatomo Nao, 'Poverty Reduction Strategy Process in Fragile States: Do the PRSPs Contribute to Post-Conflict Recovery and Peace-Building in Sierra Leone?' *Journal of International Development and Cooperation* 14:2 (2008), 67–88.

Select Bibliography

Yan Xuetong, 'Defining Peace: Peace vs. Security', *Korean Journal of Defense Analysis* 16:1 (2004), 201–19.

Yee, Albert S., 'The Causal Effects of Ideas on Policies', *International Organization* 50:1 (1996), 69–108.

Zannier, Lamberto, 'Reviving the Helsinki Spirit: 40 Years of the Helsinki Final Act', *Security Community* 1 (2015), https://www.osce.org/magazine/170891.

Zartman, William I. (ed.), *Collapsed States* (Boulder, CO: Lynne Rienner, 1995).

Zeigler, Sean M., 'Competitive Alliances and Civil War Recurrence', *International Studies Quarterly* 60:1 (2016), 24–37.

Index

Note: Figures and Tables are indicated by '*f*' and '*t*' following the page numbers.

Abyei, UNISFA operation 96*t*
accountability 79–80
Accra Accord 101–2
Afghanistan 42–3
  British lack of local knowledge
    110–12, 118
  politicization of reporting 119–20
  progress assessment 56, 63–6
African Centre for the Constructive
  Resolution of Disputes
  (ACCORD) 47–9
African Union (AU) 43–4, 101–2
aid, impact on peace duration 94
Anderson, Royce 13, 21
Angola, UN missions
  politicization of reporting 120–1
  UNAVEM II, UNAVEM III,
    MONUA 96*t*, 120–1
Annan, Kofi 33n.11
Ansorg, Nadine et al. 21–2
Armed Conflict Dataset (ACD) 82–4,
  89–90, 99–101, 103
armed conflicts 15
  classification and data measures
    16*f*, 15–19, 82–3
  definition 82
  survival analysis 84–7, 87*f*
Arusha Agreement 101–2, 102*t*
Autesserre, Séverine 107–8, 111–12
autocracy 32–4

Balkans
  contextual knowledge 112–13, 118
  data collection 55–6
  peace stability 26
Barnett, Michael et al. 38–9
benchmarking 52–5, 58, 104,
  123–5
  Burundi mission 59, 116–18
  impact of judiciary system 117
  and local perspectives 117–18
  problems with 59
  Sierra Leone mission
    (UNAMSIL) 57–8
Bhuta, Nehal 71–2

Bosnia and Herzegovina
  progress assessment 42–3, 51–2, 63,
    68–9
  UN mission (UNMIBH) 57n.16, 96*t*
Boutros-Ghali, Boutros
  An Agenda for Peace 11–12
Bozizé, François 1–2
Burma (Myanmar) 82–3
Burridge, Rupert 120–1
Burundi 19–20, 47–9, 83
  benchmarking progress 59, 116–18
  co-assessment operations 67–8
  economic growth 93–4
  Monitoring and Tracking Mechanism
    of the Strategic Framework for
    Peacebuilding 116
  Palipehutu rebellion 82–3
  UN mission (ONUB) 96*t*,
    101–2, 102*t*

Call, Charles 5–6
ceasefires 84*t*, 86, 89–90, 90*t*, 100–1,
  120–1
  in Angola 120–1
  in El Salvador 100–1
  impact on post-conflict
    rehabilitation 83–4, 90–1, 99
  see also peace agreements/settlements
Central African Republic (CAR) 1–2
Central America 17–18
Chandler, David 28
China 125
civil society 37–8, 41, 46n.50, 67, 73–4,
  101–3
civil wars
  in the Central African Republic
    (CAR) 1–2
  conflict relapse 2
  and negative peace 14–15
  onset and recurrence 3*t*, 5
  use of the term 11
co-assessment 67–8, 123–4
cold peace *see* negative peace
Cold War (-post) 11, 14–15, 40, 43,
  79–80, 84, 89n.24

146 *Index*

Collier, Paul et al. 32–3, 94, 105–6
Colombia 82–3
colonial wars 16–17
comprehensive peace/security 13–14,
  21, 35–6, 46
  by ECOWAS 44–5
  by NATO 41–2
  by OSCE 13–14, 39–40
  by the UN 37
conflict analysis 60–1, 123–5
  absence of, in East Timor 114
  environmental risks 115
  in Macedonia 61–2
  and strategic reassessment 115–16
  in Ukraine 62
contextual knowledge 106–10, 114n.29,
  124–5
Correlates of War (COW) 15–18, 68–9
Côte d'Ivoire, UN operation
  (UNOCI) 96t
Cox proportional hazards model 87–8,
  100
Croatia, UN operation (UNCRO) 96t

Dallaire, General Roméo 121–2
data collection 91–3, 113
  Armed Conflict Dataset (ACD) 82–4,
    89–90, 99–101, 103
  census data 55–6
  reliability of 74–5, 93, 118–19
  Uppsala Conflict Data Program
    (UCDP) 15–18, 68–9, 82
Davis, Kevin et al. 70
Dayton Peace Accord 51–2
Democratic Republic of Congo 11, 22,
  32–3, 118–19
democratization
  elections, and conflict outbreak 19–20
  liberal norms and values 32–4
  and peace consolidation 21–2
Denmark 73
disarmament, demobilization, and
  reintegration (DDR) 97, 100
Djibouti 73
donor states 1, 13, 45–7, 53–4, 125–6
  assistance by 94, 109–10, 117–18
  and progress assessment 2–3, 56

early warning 9, 44, 60–2, 70–1, 123–4
Eastern Slavonia, Baranja and Western
  Sirmium 105n.2
  UN mission (UNTAES) 96t

East Timor 83
  2006 violence 108–9
  conflict assessment, ethnographic
    approach 108–10
  lack of local knowledge 114, 118
  UN missions and peace settlements
    (UNMIT, UNTAET, UNMISET,
    INTERFET) 96t, 101–2, 102t,
    108–10
Economic Community of West African
  States (ECOWAS) 44–5
El Salvador
  UN mission (ONUSAL) 96t, 100–2,
    102t
elections 9, 19–20, 32–3, 58, 61–2, 94,
  98, 103, 120–1
empowerment, and resilience 28
ethnic conflicts 19–20, 60
  analysis on peace duration 91–3, 92t,
    99–100
ethnographic approach 9–10, 107–8,
  122
  strategic assessment and local
    knowledge 108–13
European Commission 61–2
Everyday Peace Indicators 75–6, 117
extra-state or extra-systemic wars 16–17

failure 1–3, 41, 77, 81, 85t, 86, 90t, 92t,
  95t, 104–5, 123
  in East Timor 83
FARC rebellion (Colombia) 82–3
Farrell, Theo 110–11
fatalities, thresholds 17–18
Finland 69–70, 73
Firchow, Pamina 75–6
Fortna, Virginia Page 79–80, 100
Fragile States Index (FSI) 69–71

g7+ 73–4
Gallie, W. B. 10
Galtung, Johan 14–15, 18–20, 48
Georgia 60
  OSCE early warning 60
  UN mission (UNOMIG) 96t
Global Peace Index 72–3
Good Friday Agreement 25
Goulding, Marrack 120–1
Gourevitch, Philip 121–2
Guatemala, UN mission
  (MINUGUA) 96t
Guelke, Adrian 51

Index

Haiti 28
UN missions (UNMIH, UNSMIH, UNTMIH, MIPONUH, MINUSTAH) 96*t*
Hartzell, Caroline et al. 79
Hastings, Max 113
Hegre, Håvard et al. 33n.9, 100–1
Helsinki Final Act 39–40
High-Level Independent Panel on UN Peace Operations (2015) 115–16
Höglund, Kristine and Mimmi Söderberg Kovacs 24, 29, 116–17
human rights 32–3, 35–7, 39–40, 42, 44, 54–5, 58, 67–70, 100–2, 117
humanitarian assistance 42, 44
difference of practices 31

income, impact on peace duration 88, 92*t*, 93–4, 102–3
indicators/indices 4, 9, 21–4, 68–70, 76, 123–4
Everyday Peace Indicators 75–6
Fragile States Index (FSI) 69–71
International Dialogue on Peacebuilding and Statebuilding 73–5
of normality 51–2
Positive Peace Index (PPI) 72–3
State Fragility Index (FSI) 71–2
see also benchmarking; conflict analysis
inequality, vertical and horizontal 5, 94, 100, 102–3
Institute for Economics and Peace 72–3
internal armed conflicts *see* civil wars
International Alert 47–8
International Committee of the Red Cross (ICRC) 31
International Dialogue on Peacebuilding and Statebuilding 73–5
internationalized internal armed conflicts 16–17
International Monetary Fund 45–6
International Peace Institute 94–7
International Security Assistance Force (ISAF) 63–6
and politicization of reporting 119–20
Strategic Assessment Capability (ISAC) 65*f*, 66, 119–20
interstate wars 16–17

intra-state or internal conflicts 16–17, 22
thresholds of violence 17–18
Iraq 42–3, 111n.18

Journal of Peace Research 14–15

Kabila, Joseph 32–3
Kiir, Salva 32–3
King, Iain and Whit Mason 112–13
Klein, James et al. 22–6
knowledge, contextual *see* contextual knowledge
forensic 112–13
generalizable 111–12
intelligence 112–13
local 108–11, 117–18
Kosovo
2004 violence 112–13
forensic knowledge and intelligence in 112–13
lack of local knowledge 118
lack of reliable data 55–6
stability in 26
UN mission (UNMIK) 96*t*
Kosovo Liberation Army 118
Kreutz, Joakim 83–4
Kurtz, Jon 46–7
Kyrgyzstan 60

Lacan, Jacques 39
Landgren, Karin 59
Lansford, Jennifer E. and Kenneth A. Dodge 18n.15
Lebanon, UN mission (UNIFIL) 96*t*
liberal democratic norms 32–4
liberal peacebuilding paradigm 35–6
Liberia 4, 48–9, 57
peace indicators 73
UN mission (UNMIL) 4, 96*t*, 101–2, 102*t*
Linz, Juan J. and Alfred Stepan 21–2
Lomé and Abuja peace agreements 57

Macedonia
census data 55–6
OSCE conflict analysis 61–2
and third-party engagement 125
UN preventive peacekeeping operation (UNPREDEP) 125
Mali, UN mission (MINUSMA) 96*t*
McGovern, Mike 77, 81

Index

McNamara, Robert (US secretary of defense) 110–11
Médecins sans Frontières (Doctors without Borders) 31
Menkhaus, Ken 28, 118–19
military victories *see* victories (military)
Monteiro, Antonio 120–1
Mozambique 73
  economic growth 93–4
  UN operation (ONUMOZ) 96*t*

negative peace 8, 18, 21, 22–4, 23*t*, 72–3, 81–2, 105
  criticisms of concept 18–19
  as minimal conception of peace 15–16
  vs. positive peace 14–15
  and security 27
Nepal 101–2
Network of Early Warning Focal Points (OSCE) 60–1
neutrality, and humanitarian assistance 31
New Deal for Engagement in Fragile States 73–4
New Institutionalism 71–2
NGOs 3–4, 30, 47, 67
  ACCORD 47–9
  International Alert 47–8
Nigeria 82–3
Nilsson, Desirée 79
non-state wars 11, 16–18
North Atlantic Treaty Organization (NATO) 30, 112, 119
  definition of peacebuilding 41–2
  progress assessment 51–2, 63–6, 119–20
  stability operations 42–3
Northern Ireland 25

Office of Internal Oversight Services (UN) 54–5
Ofstad, Olav 114
'one-sided violence' 17–18
Organization for Security and Co-operation in Europe (OSCE) 13–14, 30, 39, 54
  and comprehensive security 39–40
  conflict analysis 60–2
  constraints 41
  and post-conflict rehabilitation 40
Oslo Peace Accord 24

Palipehutu (Burundi) 67–8, 82–3
Paris, Roland 33–4
peace, peacebuilding, peacekeeping 1, 8–9
  comprehensive peace/security 13–14, 21, 37–42, 44–5
  conceptualization 8–14, 25–6, 29–30, 34, 36, 41–2, 104–5
  and consolidated democracy 21–2
  contested peace 10, 13, 24
  as a continuum 21–2
  empirical dimension 105–6
  ethnographic approach 107–13
  expenses 1
  fearful peace 25
  identification of conflict factors 5–7
  insecure peace 24–5
  normative dimension 106–7
  partial peace 24–5
  peace scale 22–4, 23*t*
  'Peace Triangle' 24–5
  polarized peace 25
  and practice 30–1
  regional peace 24–5
  and resilience 27–8
  restored peace 24
  and security 26–7
  and state-building 19–20
  and societal levels 22
  sustainable peace 28, 104
  'sustaining peace' 11–12
  unjust peace 25
  unresolved peace 24
  violence distribution 18, 104–5
  see also negative peace; positive peace; stable peace; strategic assessment
Peace and Security Council (PSC) 43–4
Peace and Security Goals 73–4
Peace Research Institute, Oslo (PRIO) 15–16
peace agreements/settlements 24, 81, 83–4
  Lomé and Abuja 57
  and peace duration 79–80, 86, 89–91, 99–102
point de capiton 39
political dimension
  as a constraint on OSCE 41
  UN activities in 37
  World Bank activities in 67–8
politicization
  in East Timor 109–10

Index

of metrics design 119
of reporting 119–21
positive peace 8, 22–4, 23*t*, 26, 35–6, 41
  and comprehensive security 39–40
  conceptualization of 19–20
  vs. negative peace 14–15
  by NGOs 47–9
Positive Peace Index (PPI) 72–3
Post-Conflict Performance Indicators
  (PCPI) 51–2
post-conflict reconstruction and
  development (PCRD) 43–4
post-conflict rehabilitation 2, 9, 11–12,
  20–1, 102–3
  by African Union 43–4
  and ceasefires 83–4, 90–1, 99
  and comprehensive security 13–14
  definition of post-conflict 81–2
  hazard model 87–98
  impact of income 92*t*, 93–4, 95*t*
  by OSCE 40
  survival analysis 77–9, 84–7
  territorial and ethnic conflicts 91–3,
    92*t*, 99–100
  and vertical and horizontal
    inequality 94, 100
  and victories 83–4, 90–1, 99
  by the World Bank 45–6
poverty reduction 37–8, 45
  Interim-Poverty Reduction Strategy
    Paper (I-PRSP) 46
power 48, 111–12
  power-sharing arrangements
    79–80
Prevlaka, UN mission (UNMOP) 96*t*
progress assessment 51–2, 76
  benchmarking 57–9
  co-assessment 67–8
  conflict analysis 60–2
  data collection 55–6
  donor fragmentation 56
  indicators 68–76
  NATO operations 63–6
  reporting 53–4
  result-based management and
    budgeting 54–5
  see also strategic assessment

Rees, Edward 109–10
reporting 3, 9, 53–4, 57–9
  and donor fragmentation 56
  by OSCE 61–2

and politicization 9–10, 119–21
  by the UN 53–4
resilience 5, 27–9, 48, 72–3
  and empowerment 28
  World Bank strategies 46–7
results-based management and
  budgeting 54–5
Richards, Paul 21
Rudloff, Peter and Michael J.
  Findley 79–80
Rwanda
  disregard of warning signs 121–2
  economic growth 93–4
  UN mission (UNAMIR) 96*t*, 121–2

Scambary, James 109–10
Schroden, Jonathan 64–6
security 3–4, 6–7, 26–9, 73–4, 124–5
  by African organizations 43–5
  comprehensive 13–14
  insecure peace 24–5
  by NATO 63–6, 119–20
  by OSCE 39–40, 62
  pluralistic security community 22–4,
    23*t*
  and post-conflict rehabilitation 13–14
  by the UN 37, 108–10
Serbia 112–13
Sierra Leone 74–5
  benchmarking progress 57–8
  economic growth 93–4
  poverty reduction 46
  UN missions (UNOMSIL,
    UNAMSIL) 96*t*, 102*t*
South Sudan 32–3
  peace indicators 69–70
  UN mission 38–9
stability operations/stabilization 11–12,
  42–3
stable peace 1, 3–4, 14–15, 19–20, 28–9,
  104, 115, 123
  factors 48, 105–6
  indicators 45–6
  concept of 10–11, 25–6
  and positive peace 19
State Fragility Index (SFI) 71–2
Stave, Svein Erik 71–2, 116n.34
Stewart, Rory (MP, UK) 110–12
strategic assessment 2–3, 56, 76, 125–6
  co-assessment 67–8
  and conflict analysis 113–15
  and contextual knowledge 124–5

Index

strategic assessment (*cont.*)
  diversity of 123–4
  ethnographic approach to 108–11
  and inter-agency competition 38
  and local perspectives 117–18, 124–5
  by NATO 52, 63–6
  and normative environment 33–4
  and peace concepts 8–9, 13–14, 34
  politicization of design and
    reporting 119
  and resilience 46–7
  and third-party engagement 124–5
  see also benchmarking; conflict
    analysis; indicators
Syrian civil war 16–17

Tajikistan, UN mission (UNMOT) 96*t*
third-party engagement 4, 11–12,
  49–50, 122, 124–5
  and co-assessment 68
  security guarantees 6–7, 79–81

Uganda 46–7
Ukraine, OSCE reporting 62
United Kingdom
  lack of contextual knowledge 110–12,
    118
  view of stability 42
United Nations Peacekeeping
  Operations (UNPKOs) 79–80,
  86–9, 94–8, 95*t*, 100–2, 102*t*, 103
United Nations Protection Force
  (UNPROFOR) 96*t*
United Nations (UN) 11–12, 30, 36, 54,
  96*t*, 102*t*
  benchmarking practices 52, 54–5,
    57–9, 116
  in Central African Republic (CAR)
    1–2
  and co-assessment 67–8
  expenses 1
  flexible conceptualization of peace 36
  impact on peace duration 79–80,
    86–9, 94–8, 95*t*, 100–2, 103
  inter-agency competition 38
  internal disagreement on
    methods 37–8
  objectives 37

reporting 53–4
in South Sudan 38–9
strategic assessment analysis 115–16
on sustainable peace 104
UN Department of Peacekeeping
  Operations (DPKO) 36, 86–7,
  94–7, 101–2
UN Department of Political Affairs
  (DPA) 36
  South Sudan mission 38–9
UN Peacebuilding Commission
  (PBC) 1–2, 36, 67, 116
UN Peacebuilding Fund 67–8
Uppsala Conflict Data Program/Peace
  Research Institute, Oslo (UCDP/
  PRIO) 15–18, 68–9, 82
US Agency for International
  Development (USAID) 27n.38,
  109–10
US Department of Defense 42
US Government Accountability
  Office 121

victories (military) 83–4, 84*t*, 86, 86*f*,
  89–91, 90*t*, 98n.39, 99, 102–3
Vietnam
  lack of local knowledge 110–11
violence threshold 12, 17–21, 81–2,
  104–5

Walter, Barbara 79–80, 94, 100
Weber, Max 30–1
Westendorf, Jasmine-Kim 41, 111–12
Westerwinter, Oliver 38
Woodward, Susan 6, 45–6
World Bank 2, 30, 67–8
  activities 45–6
  lack of data 55–6
  post-conflict engagement in Sierra
    Leone 46
  progress assessment 51–2
  and resilience-building 27–8, 46–7
World Development Report 2011 27–8
World Health Organisation 18–19
Wucherpfennig, Julian et al. 91–3,
  99–100

Zannier, Lamberto 39–40

Printed and bound by CPI Group (UK) Ltd, Croydon, CR0 4YY